IFIP Advances in Information and Communication Technology

357

T0223411

IFIP – The International Federation for Information Processing

IFIP was founded in 1960 under the auspices of UNESCO, following the First World Computer Congress held in Paris the previous year. An umbrella organization for societies working in information processing, IFIP's aim is two-fold: to support information processing within ist member countries and to encourage technology transfer to developing nations. As ist mission statement clearly states,

> *IFIP's mission is to be the leading, truly international, apolitical organization which encourages and assists in the development, exploitation and application of information technology for the bene t of all people.*

IFIP is a non-profitmaking organization, run almost solely by 2500 volunteers. It operates through a number of technical committees, which organize events and publications. IFIP's events range from an international congress to local seminars, but the most important are:

- The IFIP World Computer Congress, held every second year;
- Open conferences;
- Working conferences.

The flagship event is the IFIP World Computer Congress, at which both invited and contributed papers are presented. Contributed papers are rigorously refereed and the rejection rate is high.

As with the Congress, participation in the open conferences is open to all and papers may be invited or submitted. Again, submitted papers are stringently refereed.

The working conferences are structured differently. They are usually run by a working group and attendance is small and by invitation only. Their purpose is to create an atmosphere conducive to innovation and development. Refereeing is less rigorous and papers are subjected to extensive group discussion.

Publications arising from IFIP events vary. The papers presented at the IFIP World Computer Congress and at open conferences are published as conference proceedings, while the results of the working conferences are often published as collections of selected and edited papers.

Any national society whose primary activity is in information may apply to become a full member of IFIP, although full membership is restricted to one society per country. Full members are entitled to vote at the annual General Assembly, National societies preferring a less committed involvement may apply for associate or corresponding membership. Associate members enjoy the same benefits as full members, but without voting rights. Corresponding members are not represented in IFIP bodies. Affiliated membership is open to non-national societies, and individual and honorary membership schemes are also offered.

John Impagliazzo   Eduard Proydakov (Eds.)

Perspectives on Soviet and Russian Computing

First IFIP WG 9.7 Conference, SoRuCom 2006
Petrozavodsk, Russia, July 3-7, 2006
Revised Selected Papers

 Springer

Volume Editors

John Impagliazzo
Hofstra University
Fort Salonga, NY 11768-2710, USA
E-mail: john.impagliazzo@hofstra.edu

Eduard Proydakov
Virtual Computer Museum
Moscow, 123060, Russian Federation
E-mail: e.proydakov@yandex.ru

ISSN 1868-4238   ISSN 1868-422X (eBook)
ISBN 978-3-642-27082-6   ISBN 978-3-642-22816-2 (eBook)
DOI 10.1007/978-3-642-22816-2
Springer Heidelberg Dordrecht London New York

CR Subject Classification (1998): K.2, B, C, D, E, F, G

Typesetting: Camera-ready by author, data conversion by Scientific Publishing Services, Chennai, India

Printed on acid-free paper

Springer is part of Springer Science+Business Media (www.springer.com)

Dedication

We dedicate this book to

the men and women who seek to preserve the legacy of the computing profession, particularly those from the former Soviet Union, whose accomplishments and dedication to computing have propelled the work of their colleagues and have enhanced the computing profession for future generations.

Foreword

Eduard Proydakov, the person who has accomplished much to reconstruct the history of Soviet computer technology, asked me to write a foreword for this collection of papers on some historical Soviet projects. Notably, the authors themselves are the participators of these projects and, generally speaking, historical persons as well. I could not refuse Eduard's request but immediately faced an ethical problem. Do I have the right to write about this with the sincerity it should have, and how would the authors of these articles perceive this sincerity?

It fell upon me to work with outstanding people or to meet them. Some of them have died; others are among the authors of this collection of articles. Therefore, I decided to write about my own perception of the professional world in which I lived and continue to live. Certainly, this is a small piece of a larger computer world. However, I perceive it has a vast and instructive value for me, and I would like to write about my own perception of this value.

I'll start from university. Some time ago I was privileged to hear a very interesting talk that explained why our generation preferred to choose the sciences (e.g., physics, mathematics, ...) rather than humanities (e.g., philosophy, history, ...). The reason was an internal freedom, the freedom from falsehood of the communist ideology. You do real work, and you do not need to relate this work with an ideology. Obviously, I could not formulate this when I was 17 years old, but now I think it was the main reason I chose the Moscow Institute of Physics and Technology (MIPT) for my studies. Another important reason was that graduates of MIPT worked in research institutes at the great Academy. The 2010 Nobel laureate in physics, Konstantin Novoselov, had also graduated from MIPT. He said in one of his interviews that no university of the world (including universities of the USA and UK) provided a better level of education than MIPT. I do not believe that he is right, but in any case, MIPT is an excellent technical university. At that time, MIPT had several peculiarities. First, there was a very heavy academic load, especially on physics and mathematics. The total number of academic hours (lectures, seminars, laboratory training) during the first two years of education was at least 52 hours per week. As a result, students received a very broad education. It was not strange that my classmates became specialists in various areas such as in radio engineering, pure mathematics, and theoretical physics. I became a programmer.

The second peculiarity of MIPT was the model of education. Starting from the fourth year, we spent a part of each week at so-called base educational departments within the walls of their respective research institutes where we had lectures and were gradually beginning to participate in real projects as junior colleagues of a research staff. My base educational department was at the Institute of Precise Mechanics and Computer Engineering (IPMCE). The director of IPMCE was the great academician Sergey Alexeyevich Lebedev. I still remember my first impressions from lectures at Lebedev's educational department. At MIPT, they provided us with stable, traditional, and well-prepared lectures; IPMCE offered us some kind of improvisation. It was not clear what lecturers (even Lebedev) were talking about and why. Later, we certainly had adapted ourselves to such a manner and understood that lecturers saw us as their future colleagues, and they talked to us about problems that they needed to solve. As computer engineering was a very young field, solving each problem was a discovery, and lecturers of the base department told us about their discoveries and about their happiness to explore these new things. In addition, they gave us much freedom to choose a direction of work. I had changed two scientific advisors for different reasons until the sixth year, when I started to work in Lev Nikolaevitch Korolev's laboratory of programming. At that time, I considered programming as a profession for elderly women (30–35 years old; please note that then I was a little over 20 years old). I thought a man should do some other work. However, Korolev provided me with half-pay that was equal to my student grant, and this was the main reason to work as a programmer. Only later during my diploma preparation did I realize the amazing attraction of the profession of a programmer.

After graduating from MIPT, I continued to work at Korolev's laboratory. S.A. Lebedev created at his institute an amazing professional and humanistic climate. There was absolute openness and kindness among members of staff. I was able to go up to any person whose work was interesting for me and I always received a well-intentioned and exhaustive explanation. Sometimes, even in a smoking room I would suddenly think of a solution for a task that I could not solve for several days. We young folk perceived such an atmosphere as the norm and we tried to follow these established practices. (By the way, an organic perception of this humanistic and professional culture came to me only over time.) We often got together with older friends outside the institute: drinking vodka, visiting ice rinks and swimming pools, and attending hockey and football matches. Some conflicts probably existed, but they were at higher organizational levels and they did not affect the staff members or us. Once again, we had great freedom to choose a topic of research. Here is one example. We were making preparations together with engineers for tests of the BESM-6. I was responsible for several large pieces of the D-68 operating system and time was very tight. However, I had become interested in debugging and was developing a multiuser interactive debugger; it looked like a time-sharing system. I was writing and debugging the system mostly on non-working days. I do not think that Korolev was delighted with my immersion but he did not exhibit any discontent even in a subtle manner. It seemed that our life was determined forever. Engineers designed another machine (certainly, we participated in a design of its architecture) and we, programmers, developed all the system programs for it. This was perfect. I wished to live this way forever.

Just after graduation from MIPT, I started to lecture in various universities. Themes of lectures did not confuse me: programming in Algol-60, computer architecture, operating systems, compilers, databases; one time I even lectured about the IBM/360 for a full year. Over time I had my own students and relations with them were based on Lebedev's practices. Then my own subordinates appeared, and it became possible to perform larger projects. Again, despite the fact that I lectured in working hours, my boss Korolev only welcomed this activity. I should note that generally many staff members were teaching at universities. It was the norm. Today we call it an "integration of sciences and higher education". Many years later, I realized that one kind of output of the academy (probably the most important one) was humans, experts with the highest qualification.

Here is one more very funny illustration of the openness that predominated. Within the institute, two groups designed competing computers: the group of Melnikov and the group of Burtsev. These heads did not like one another too much but we, as members of these groups, openly exchanged ideas and any available information. Certainly, the bosses knew about this but they did not try to stop it.

The circle of my acquaintances gradually went beyond the boundaries of the institute. These new places included Academgorodok (Novosibirsk), Kiev, the Computing Center and the Institute of Applied Mathematics at Moscow, and the Spaceflight Control Centers. Again, I met with the same openness, the same kindness! What clear minds, what famous names! There were many conferences and winter schools on various specific topics.

Of some discomfort was the access to foreign journals and magazines; access was difficult, with significant delay and always scarce. Certainly, the openness within our informal professional society was a great help. We always exchanged fresh interesting papers. Andrey Petrovitch Ershov played a great part in this scene; he organized international conferences and workshops with the participation of leading Western scientists. We were always interchanging fresh interesting articles. At that time, I did not have personal contacts with Western colleagues because I had access to secret information as early as at the second year of education. Moreover, it was only in 1984, after my election as a corresponding member of the Academy of Sciences, that I had the possibility to communicate and to participate in foreign conferences.

I turned down many proposals to change my job. The job at IMPCE was a fantastically interesting job and the environment was very comfortable. However, the time had come in 1980 to leave IPMCE. It happened as follows. V.A. Melnikov had left the institute and joined one of the institutes of the Ministry of Electronic Industry. All my projects related to computers were designed by him. I promised to join him and I did. Some key engineers of Melnikov's laboratory (my very intimate friends) had decided to stay at IPMCE; I was incredibly sad about this. I remember that for many years I physically perceived a strong nostalgia when I passed by the institute. In addition, I had learned that another, rather uncomfortable, climate existed. The environment was depressive and it was not always possible to detach yourself from it. One more circumstance was unpleasant: administrative overloading. I became a head of two departments – the department of system programming (it was normal) and the department of computer CAD/CAM – with 150 employees in total.

Many years have passed since then. During the last 17 years, I have directed the Institute of System Programming. During this time, our country has changed radically. Values have changed; moral criteria transformed wildly. All relationships between industry, education, and science have been destroyed. The industry of high technology is significantly disrupted. The academy and universities are degrading. It seems sometimes that this state of affairs is improving; perhaps this perception is a result of expectations and wishes for improvement. Yes, something has changed, for instance, the brain drain has decreased. However, these days are generally not the best for the academy and universities. The Soviet social respect of scientists and professors was eventually diluted. Government and society barely demand science. Western multinational corporations dominate the domestic market of information technology. This is the today's reality.

At the same time, the values that I adopted at IPMCE, the values that had become my own values, not only continue to exist for me but have become even more important. I am probably too categorical but I cannot believe that our country has worthy chances for the future without the academy, without technical professionals, or without their high professional and humanistic ethics. I really believe that all these values, the academy, and the professionals will be in demand again. In addition, this is not just a passive expectation; my colleagues and I carry on serious practical activities within the institute and at universities. In the institute, more than 100 undergraduates and doctoral students take part in projects together with their senior colleagues. Certainly, a significant part of these projects consists of outsourcing, which we may consider industrial research, that is, the development of new technologies. I hope that fraction of outsourcing will decrease with time thanks to an increase in domestic orders for the development of new technology. As it happens, we may efficiently use the model of IPMCE – composition of basic research, development of new technologies, and education – in the new environment also. Talking about the current new generation, this generation seems to be better, more talented than my generation. I enjoy learning about their results, their articles, their new courses of lectures. I wish to believe they have a future here in our country.

So it goes. However, this is only one part of the landscape. In Soviet times, an industry of programming did not exist as an independent branch of industry. However, in post-Soviet times this industry was formed. It is very important that the formation of this industry in Russia occurred without any government control and participation. Certainly, this industry has not appeared from scratch. People from the defense industry, the academy, and the universities generally founded these companies. Many successful non-government companies work in the market of software products and services not only in our two capitals but also in the provinces. Some of these companies are high-technological enterprises that successfully compete in the world market. There are very interesting processes going on in this branch of industry. Several associations have been established that involve private and government organizations. These associations form general strategies of the industry and successfully lobby its interests through government structures. A serious business cooperation of these associations has developed. The first sprouts of interaction between the industry, the academy, and education have appeared. All this holds forth a hope to form a sound ecosystem. I would like to conclude this foreword on this optimistic note.

Victor P. Ivannikov
Academician
Institute of System Programming
Russian Academy of Sciences
ivan@ispras.ru

Preface

These proceedings represent a collection of papers derived from the first conference on Perspectives on Soviet and Russian Computing, also known as *SoRuCom*. The conference took place in July of 2006 in Petrozavodsk, Karelia, Russia. Over 170 participants from the former Soviet Union, the United States, the European Union, and Cuba attended the event.

These proceedings reflect much of the shining history of computing activities within the former Soviet Union. Despite the existence of several publications in this history, they often tarnished reality. The primary reason for this blemish was the mode of privacy and secrecy that existed in the former Soviet Union during the early days of computing. Indeed, many archival materials were lost in the so-called reorganization of the former union. Additionally, it is important to note that the government during that time had given little value to the preservation of historical artifacts.

In recent years, however, this position gradually began to change for the better. Now, many new museums exist at private and educational institutions. For example, for the last 15 years, Eduard Proydakov's Virtual Computer Museum (www.computer-museum.ru) has collected and preserved numerous documents such as material on the history of computing, material on outstanding domestic scientists and engineers, and the memoirs of computer developers and pioneers.

* * * * * *

The preparation of this book has taken a very long time. The basic difficulty was that the vast majority of its authors do not speak English and painstaking translations and transcript verification was necessary. Moreover, in recent years, computer facilities and terminologies have changed. Each change of new devices resulted in some rewriting that may have modified the intent or the meaning of terms. Here it is necessary to express words of gratitude to John Impagliazzo, who has devoted much time and work to edit the articles in this book.

* * * * * *

In looking at computing history in the USSR, it is possible to identify six stages of its development. We explain these periods in the following discussion.

1. Origins (1948–1960) – The Creation of the First Lamp Computers

As with the United States in the 1950s, many people in the former Soviet Union believed that the country needed only ten computers. Therefore, the country delivered only seven "Arrow" machines, the first serial computer in the country. Its operation was under the personal control of the head of the state, I.V. Stalin, who spoke about the value received from the given work. They used the first computers for military decision-making problems and for research purposes.

Therefore, in 1956 in Moscow there appeared the first machine translation systems. It is clear that it was an attempt to solve a problem at different levels of complexity, but it has been shown that the given problem was very difficult and was accomplished only in Russia on the basis of artificial intelligence techniques; today's systems provide comprehensible translation quality. By the end of the 1950s, it became clear that computers could solve a much wider range of problems. Hence, it was necessary to organize the manufacturing industry to develop and expand professional training for this new machine.

2. Origin of the Computer Industry (1961–1970)

The development of serial computers and the involvement of computers as a part of the arms systems, particularly in air defense systems, characterizes this period. The 1960s showed the beginnings of the expansion of manufacturing factories for computers and the new constructions in Minsk, Kazan, and other cities. The country initiated an enormous program to develop a robust semiconductor industry. Due to the conditions of the cold war, the USSR was compelled to undertake these branches independently. At that time in the USSR, we saw the development of new original architectures for computers such as N. Brusentsov's figurative computer and the computers in residual classes by I. Akushsky and D. Yuditskiy.

3. The Copying Period: Formation of the Computer Industry (1970–1980)

The manufacture of computers in the former Soviet Union was gaining momentum. However, by the mid-1960s it became clear there was a build-up from the West. While the introduction of computers in military systems, technological equipment, and scientific research developed quite well, their introduction into the economy was greatly hampered by excessive regulation and lack of internal process incentives. To reduce the gap with the West, the government decided to release computers that were compatible with IBM mainframes and with DEC minicomputers for industry. This decision had both positive and negative consequences. Positive aspects consisted of cooperation in the manufacture of computers with countries of Eastern Europe; between them, there was a division of labor in the field. They began issuing computers in the tens of thousands a year as well as organizing the preparation and retraining of personnel. Negative consequences consisted of the companies financing their own work; funds were severely lacking and, as a result, the situation did not provide the possibility to develop breakthroughs and strategic directions in the field of computer architecture. The situation had placed the country in a position of constantly trying to catch up with the United States.

4. The Beginning of the Microprocessor Era (1980–1990)

Small companies worked out their own problems. The appearance of the first domestic eight-bit microprocessors had provided an opportunity to develop the microcomputer within the limits of the small companies, institutes, and enthusiasts. Personal computers of different types became very popular and were in short supply. However, the more advanced companies and institutes bought personal computers both from the Soviet manufacturers and from imports. Neverthe-less, the situation allowed the large state computer centers to gain in strength, becoming compatible with Western centers; however, they often modified the machines to have various accelerators for new types of computers developed for the solution of special problems. This was a time of the greatest achievements of Soviet science and technology.

5. Crises of the 1990s

The restructuring of the former Soviet Union and the withdrawal pains from a new social order in the field of information technology caused stagnation in technical development, except for software development. Russia began to use imported equipment. However, private information and communication technol-ogy (ICT) companies actively engaged in system integration development. They began with the development and installation of simple local networks that even-tually led to the construction of large-scale network systems. This became the dawn of a full integration into a worldwide system that would utilize segmented efforts of labor.

6. The Modern Period

Russian ICT grew substantially, especially in the field of custom-made program-ming; the developments became part of a global system of labor. Companies conducted their own work in small volume. However, there was a huge shortage because millions of experts in the field of high technologies departed abroad, which caused many economic problems. Nevertheless, in a declaration by D.A. Medvedev, the president of Russia, a new modernization of the Russian economy led to one of the five basic directions, namely, the development of ICT. Efforts were carried out through the center of innovative development of "Skolkovo" in cooperation with the leading Western ICT companies such as Intel, Microsoft, Cisco, and Nokia.

* * * * * *

We now express our gratitude to all those individuals who made the SoRu-Com conference of 2006 a reality. They include the authors of the papers contained within these proceedings, the presenters, the participants, the So-viet computer pioneers, the students of those pioneers, and the organizations (acknowledged herein) that provided financial support to make the event pos-sible. In particular, we thank Ruzanova Nataliya Sokratovna, Vice-Rector of Petrozavodsk University, for her many efforts and unrelenting belief in making the conference possible. We also thank Iurii Bogoiavlenskii of Petrozavodsk Uni-versity for helping to shepherd the process to make the event come to fruition. In a modest gesture, SoRuCom recognizes John Impagliazzo for initiating and organizing the event and journalist Eduard Proydakov for soliciting many con-tributors to the conference. Indeed, SoRuCom was a memorable event.

* * * * * *

We realize that this book does not resolve the total history of the development of computing in Russia and in the former Soviet Union. However, we hope that it has captured and preserved parts of Soviet computing history and that it will be useful to historians and to readers who maintain an interest in this subject.

June 2011

John Impagliazzo
Eduard Proydakov

Institutional Organizers

International Federation for Information Processing (IFIP)
Ministry of Education and Science of Russia
Federal Agency for Education (Rosobrazovanie)
Government of Karelia
Petrozavodsk State University
National Fund for Professional Training
State Research Institution of Information Technologies
and Telecommunications "Informika"
Saint Petersburg State University
Polytechnic Museum (Moscow)
Virtual Computer Museum

Local Organizing Committee

Vasiliev, Viktor Nikolaevich	Russia, Chair
Ruzanova, Natalia	Russia, Co-chair
Impagliazzo, John	USA
Nasadkina, Olga	Russia
Shlykova, Svetlana	Russia
Golubev, Alexey	Russia, Secretary

International Program Committee

Khetagurov, Yaroslav A.	Russia, Co-chair
Prziyalkovskiy, Viktor Vladimirovich	Russia, Co-chair
Agomirzyan, Igor	Russia
Alexandridi, Tamara Minovna	Russia
Bogoiavlenskii, Iurii A.	Russia
Brusentsov, Nikolay Pevrovich	Russia
Ivannikov, V.P.	Russia
Impagliazzo, John	USA
Fitzpatrick, Anne	USA
Karpilovitch, Yury Vladimirovich	Belarus
Malinovskiy, Boris Nikolaevich	Ukraine
Oganyan, German	Armenia
Prochorov, Nikolay	Russia
Proydakov, Eduard	Russia
Rogatchov, Yury Vassilyevich	Russia

Svinarenko, A.G.	Russia
Smolevitskaya, Marina Ernestovna	Russia
Tikhonov, A.N.	Russia
Tomilin, Alexsander Nikolaevich	Russia
Voronin, Anatoly V.	Russia
Zgurovsky, Mikhail	Ukraine

Editorial Committee

Impagliazzo, John	USA, Co-editor-in-Chief
Proydakov, Eduard	Russia, Co-editor-in-Chief
Agomirzyan, Igor	Russia
Gerovitch, Slava	Russia
Nitusov, Alexander	Russia
Polyak, Yuriy	Russia
Fitzpatrick, Anne	USA

Acknowledgments

The conference was organized by the International Federation for Information Processing (IFIP) Working Group 9.7 (History of Computing), Technical Committee 9 (Relationship Between Computers and Society). The organizers of this First Conference on Perspectives on Soviet and Russian Computing (SoRuCom) wish to extend their appreciation to the following groups:

o International Federation for Information Processing (IFIP)
o Ministry of Education and Science of Russia
o Federal Agency for Education (Rosobrazovanie)
o Government of Karelia
o Petrozavodsk State University
o National Fund for Professional Training
o State Research Institution of Information Technologies and Telecommunications "Informika"
o Saint Petersburg State University
o Polytechnic Museum (Moscow)
o Virtual Computer Museum

Additionally, we thank the following donors for monetary or other support. Without them, the conference could not have occurred.

o Petrozavodsk State University
o International Federation for Information Processing (IFIP)
o Institute for Electrical and Electronic Engineers (IEEE) Foundation
o Association for Computing Machinery (ACM)
o IEEE Computer Society
o ACM Special Interest Group on Computer Science Education (SIGCSE)
o Cisco Systems
o IBM Russia
o Microsoft Russia

Table of Contents

The Work of Sergey Alekseevich Lebedev in Kiev and Its Subsequent Influence on Further Scientific Progress There

Zinoviy L. Rabinovich

V.M. Glushkov Institute of Cybernetics
Kiev, Ukraine

Abstract. The paper represents the memories of an active participant of development of the first Soviet computers in Kiev, Ukraine. The process of creating electronic computing machines such as MECM, BESM and SESM is described in detail. The outstanding role of S.A. Lebedev (head of the research group) is specially stressed as well as his impact on subsequent progress.

Keywords: MECM, first Soviet computers, Academician S.A. Lebedev.

1 Introduction

In my opinion, the day of the 6 November 1950 was the culmination of Lebedev's entire work in Kiev. This was the day when the MESM–still a prototype–solved its first test problem. That was a miracle! The electronic computing machine was created in the Ukraine!

Here, I will briefly recall what occurred before and after that moment. One should not forget that Sergei Alekseevich Lebedev already was an outstanding scientist [1] before he was appointed to the position as director of the (Kiev) Institute of Electrotechnics. The institute was established with Lebedev's participation. In the beginning, he headed the laboratory of modeling and automatic control there. Then, that laboratory transformed into laboratory of computing machinery at about the end of 1948. That was the time when the development and construction of the small electronic computing machine (MESM) started.

2 The Birth of MESM

We could characterize the works performed at the laboratory before that date as the preparation phase for the MESM machine. However, those works themselves also had significant importance, independent of their final objectives. Those were the electronic research and development (with subsequent adjustment) of a special installation for semi-natural modeling and testing of the aircraft flight stabilization systems. That system was made under the order of another research institute. However, the payment received for the work became sound financial backing for initial development of the MESM, which, in fact, began as an experiment launched

J. Impagliazzo and E. Proydakov (Eds.): SoRuCom 2006, IFIP AICT 357, pp. 1–5, 2011.

entirely on the laboratory scientists' own initiatives. Nonetheless, the idea to construct this installation was a notable scientific achievement of Sergey Alekseevich. Authoritative scientific sources subsequently highlighted that accomplishment.

Actually, among its other activities, the installation integrated in its structure a special analog computing unit (in fact an analog computer) based on operation amplifiers; it modeled the movements of an aircraft during its flight and provided a dynamic platform that operated with three degrees of freedom. The latter had modeled real angles of movements and distortions. A special high-precession monitoring system, mounted between them, transformed the modeled aircraft flight distortions into power impulses or impacts applied on special torque motors of the platform.

Not only the ideas, but also their technical solutions proposed and realized by Sergey Alekseevich Lebedev were original. That was the reason why the whole project was rightfully named, 'the pioneer one'. I have described it here in such detailed way because few knew about it, as it was overshadowed by his own later brilliant achievements in electronic computer development. However, one should mention that this analog computer, devised by Lebedev for the system, was one of the first Soviet computing machines of the analog type.

Before beginning his practical work on the computer MESM project, Sergei Lebedev personally performed impressive scientific theoretical preparations. He found and developed basic technical solutions for design of the MESM and its principal structural circuits and structure. However, before that, he developed the fundamental general principles for the architectural creation of digital electronic computers—an entirely new kind of computing machinery. In other words, the principles, which are famous as the John von Neumann architectural principles, Lebedev discovered and formulated independently; that is, he invented digital electronic computer all by himself. Construction of the computer lasted two years until the first test run, and one more year until the official governmental commission accepted it and put it into regular operation on 25 December of 1951.

3 From MESM to BESM

Rapid performance, considered a necessity to work with absolutely new type of (computing) machinery, became real not only due to Lebedev's own thorough scientific preparations, but almost equally, because of a special project work schedule, or rather a *style* of the project team's work, introduced by Lebedev and L.N. Dashevskiy during his first scientific deputy at the laboratory. Thus, most of them almost permanently lived in the building of the (secret) research centre, a little off Kiev, where they were assembling the computer. Lebedev himself and his chief engineer P.J. Chernyak lived there with their families. Therefore, those circumstances as well as the unbelievable enthusiasm of all, particularly the young collaborators, became a sure guarantee for their future success.

Another remarkable, and scientifically significant fact regarding the project, was the work of A.A. Lyapunov's team on the new computer. As soon as the MESM was available to perform (although before the official submission), the USSR Academy of Science dispatched an outstanding mathematician and programmer of Moscow Aleksey A. Lyapunov with some other scientists to Lebedev's laboratory for several months to conduct mathematical experiments and perform some urgent calculations for defense related problems.

Besides general impulse, given by appearing MESM for the Soviet following computer development progress, as well as its being itself a long-awaited means for solving urgent problems, the MESM played an essential role as a fully operating model (and thus, most important research tool) for development of the following "full-scale" series BESM of "big" computers. Those machines were already in production in Moscow, where Lebedev returned in 1950 to head the Institute of Precision Mechanics and Computer Engineering (IPMCE). Later, this became the most prominent scientific research institute on computers in the USSR. His invaluable experience of the work in Kiev was one of the keys to its further success.

Indeed, the MESM computer did have all the features to receive consideration as a model for the BESM machine. Namely, its architecture already possessed a number of qualitative characteristic features for efficiently developed in later computers. Thus, the ability to perform operations with commands (programming statements) was one of them. This fact enabled Prof. B.N. Malinovskiy of Kiev (then a young assistant of S.A. Lebedev and now famous historian of computing) to define the MESM as the world's first computer with such features.

4 From BESM to SESM

Lebedev's phenomenal success in creating the MESM was not the only achievement of his Kiev laboratory. The MESM was a predecessor of the, so-called "big machines" with maximum possible parallelism of information transmission and processing. However, already at that time Sergey Lebedev was also thinking about mini-machines. Consequently, he initiated the development of the sequential arithmetic logic device/unit (ALU) through sequential information transmission and processing. The operational unit was based on dynamic registers on a magnetic drum. That is, it provided maximum economy of electronic components, some of which were not very reliable then. Later, his collaborators discovered that the unit Lebedev devised was intended for a special mathematic mini-computer called SESM, which should solve linear algebraic equations with iteration methods. Later they used this machine for various vector-matrix procedures.

Lebedev's basic idea of the SESM consisted of simultaneous data input and data processing. The sequential ALU of the SESM proved to be quick enough for that purpose. Therefore, considering its destination and considering the principle of combined parallel data input and processing, the SESM could be–and did–defined as a prototype of modern matrix-vector processors, built into the structure of high-performance universal digital electronic computers. Lebedev proposed the general design of the computer, but it was the author (Z.L. Rabinovich), under Lebedev's scientific supervision, who brought it from a detailed project to a functioning machine. Lebedev consulted the Kiev team and he liked the computer they made.

In other words, the SESM, although assembled by Lebedev's collaborators after his departure to Moscow, was part of his achievements of the Kiev period. However, Lebedev himself, being a very modest person, never accepted this idea; since he was working in Moscow then, he did not directly participate in the computer construction.

The SESM was the second computer made in the Ukraine. It became operational at organizations based on Lebedev's "Laboratory for Modeling". Those were the laboratories of the famous mathematician, Prof. B.V. Gnedenko, which later transformed into the Computer Centre of the AS Ukraine headed by V.M. Glushkov, and still later into the (Kiev) Institute of Cybernetics. The SESM was a civil, scientific oriented computer that operated freely without military secrecy restrictions. Therefore, its appearance caused a positive international resonance. Subsequently, a monograph about the computer published by its designers was translated and distributed in the USA.

Unfortunately, the SESM was the last scientific project in Kiev where Sergey Lebedev was concerned. However, we should not take this literally. After his departure to Moscow, Lebedev did not lose interest in the developments in Kiev. Periodically, he would visit and his former collaborators always asked him personally for an advice or some other assistance. Those relations also manifested themselves through the cooperation between Lebedev's IPMCE of Moscow and Glushkov's Kiev Institute of Cybernetics. These two prominent researchers created world-recognized schools of science that had developed the most advanced electronic computers for mass usage. They distinguished themselves with original progressive technical solutions for problems at the world's highest level. The world scientific community also recognized their achievements as 'Computer Pioneers' by decorating them with subsequent medals, unfortunately awarded posthumously.

The aforementioned cooperation had precisely defined activity fields and it made a very positive influence on the progress of Soviet computing. There was a mutual generation of ideas on increasing computer performance as well as on the development of the "machine intellect" [2] or machine intelligence, on machine realization of the high-level programming languages, on processing knowledge methodology, and developing the dynamical organization of machine computations.

5 The MIR and ELBRUS Machines

The second direction of the cooperation proposed by the Kiev Institute of Cybernetics, especially when it concerned realization of high-level language (HLL), was received as a "revolutionary" one. Therefore, it had caused fierce arguing in the beginning. An important supporting role of S.A. Lebedev in its development was remarkable and, to a certain extent, decisive. Moreover, Lebedev developed the principle of structural interpretation of HLL; he preliminarily approved this principle in its general form and tested in details in operation of mini-computer "MIR" designed for engineering calculations and in the experimental model of the patented high performance computer "Ukraine". This principle was also implemented in the multiprocessor computation complex (super-computer) "ELBRUS", considered as an outstanding achievement of the Soviet computer engineering.

Development of the serial computer "MIR" became a similarly outstanding achievement, but in the field of small computers with advanced machine intellect. The mentioned synthesis of ideas of increasing computer performance and of the development of machine intelligence (computer "intellectualizing", so to say) is still in progress. Thus, the Kiev Institute of Cybernetics was developing computers with distributed information processing (parallel architecture), realization of a very high-level programming language, and advanced methods of knowledge processing (so-called an "intellectual solving machine – ISM"). This computer is designed under the supervision of Prof. V.N. Koval, who implemented it with the experience of an earlier developed 'essentially multi-processor complex' with 'multi-conveyor' organization of computation.

AS Ukraine also awarded collaborators of the both institutes (named after Glushkov in Kiev, and after Lebedev in Moscow) with Lebedev's premiums and also for joint projects. This recognition is clear evidence of the friendly relations that existed between the two institutions.

6   Conclusion

I would like to finish this narration with an historical phrase of Sergey Alekseevich Lebedev who said on the shore of the lake at Theofania (or Feofania), the Kiev suburb where he made his MESM in the secret laboratory during his visit there. He waved his hand towards the place and said,

"Here, it was here where we started all this".

It would be wonderful if *all this* could receive further progress, not less intensive then at those times.

End Notes

1. Soon after the war, Professor Dr. of Technical Sciences, S.A. Lebedev (1902-1974) of Moscow, was elected to the Academy of Sciences Ukraine (AS Ukraine) and he was invited as an experienced specialist to Kiev to participate in the post-war restoration and further development of the science in the republic.

    Editors' Note to #1
    That was a part of the USSR All-Union program of the post-war development.

2. «Machine intellect» - (degree of) technical/structural ability (or perfection) of a computer to realize efficiently algorithms and programmed tasks, including those composed for «artificial intellect».

    Editors' Note to #2
    The author, Z.L. Rabinovich (1918–2009), was one of the first, and undoubtedly the leading, Soviet specialists in this field.

History of the Creation of BESM: The First Computer of S.A. Lebedev Institute of Precise Mechanics and Computer Engineering

Vera B. Karpova[1] and Leonid E. Karpov[2]

[1] Lebedev Institute of Precise Mechanics and Computer Engineering, Russian Academy of Sciences, Manager of the Institute History Museum
kotre@mail.ru
[2] Institute for System Programming, Russian Academy of Sciences, Senior System Analyst
mak@ispras.ru

Abstract. Some aspects of the history of the first Soviet computer development are described. The idea of this work was offered by Sergey Lebedev who formulated the principles, wrote a plan and suggested the structure of the computer which was named BESM – Fast Electronic Calculating Machine. Some new documents are presented – a working notebook with Lebedev's handwritten notes and a plan of a preliminary project draft development. BESM computer became the first in the long series of Soviet computers built under the leadership of Lebedev. It also was used as a prototype of the first Chinese computer built with the help of Soviet engineers.

Keywords: digital computer history, Darmstadt conference, Chinese computer.

1 Introduction

Institute of Precise Mechanics and Computer Engineering of the USSR Academy of Sciences (AS USSR) was established in 1948. At that time, two types of calculating devices (forerunners of modern digital electronic computers) were under development. They were the analog computers (mechanical or electronic but not digital) and the relay computers (digital but electromechanical, not electronic). Both the principles of computer design and computers themselves were introduced in the USSR by Sergey Lebedev (1902-74), shown in Figure 1. One could say that it was predetermined in the very beginning of his scientific activity. In 1933, Lebedev and Prof. Petr Zhdanov published their joint monograph "Stability of parallel functioning of electrical systems" [9]. His thesis for a doctoral degree was also devoted to this topic. Almost every project of Lebedev was built on enormous amount of mathematical calculations. Quite logically, that brought up the idea of some facility that could automate those routine operations. That idea was in his mind permanently.

To calculate parameters of the power transmission line from the river Volga to Moscow (several thousand miles) he managed to install highly automated device constructed of powerful coils and capacitors. That device mathematically modeled the real transmission line. Implementation of that model (in fact – a special calculating

J. Impagliazzo and E. Proydakov (Eds.): SoRuCom 2006, IFIP AICT 357, pp. 6–19, 2011.

device) helped to perform all the calculations and produced a complete set of project documents for that unique line. This work was partly repeated by Lebedev during the World War II when he was engaged in design of stabilizing system for tank gun and an automatic targeting device for aircraft torpedo. Those devices could perform both basic arithmetic operations, and differentiation and integration; yet they were still analog devices. Nevertheless, there are evidences that Lebedev was interested in digital computing devices long before the war.

During 1948 and 1949 Lebedev formulated basic principles of electronic calculating machine architecture that were very similar to Janosh (John) von Neumann's principles (see Figure 2), who described his conception in a secret report in 1945. This work became known in the Soviet Union much later; all of Lebedev's work (both theoretical and practical) was carried on independently. Lebedev emphasised the following requirements:

o   A computers must have arithmetic device, memory, control and input/output units;
o   Both the program written in machine codes and the numbers should be stored in the same memory;
o   Binary notation should be used for representation numbers and instructions;
o   All calculations should be performed automatically according to the (stored in memory) program;
o   Logical operations should be performedin addition to arithmetic operations;
o   Computer memory should have a hierarchic structure.

Fig. 1. S.A. Lebedev (1950s)   **Fig. 2.** Janosh (John) von Neumann

In 1945, Lebedev was elected a member of the Ukraine Academy of Sciences (AS). There he started his work on the electronic calculating machine MESM (Russian abbreviation of "Model Electronnoy Stchetnoy Mashiny" – Model of Electronic Calculating Machine), in 1947[5]. His actual intention was to begin researches with creating just an operating model but, considering the (post war) financial difficulties of the Academy, such "luxury" was impossible. However, MESM was quite operable computer. Design of storage device and other basic units (including general computer structure) was finished by the end of 1949. The first half of the 1950s was dedicated to manufacturing and adjusting of the machine components.

2 BESM and Lebedev's Technical Notes

On 16 March of 1950, Lebedev was appointed a director of the Laboratory No. 1 at the Moscow Institute of Precise Mechanics and Computer Engineering (now - IPMCE AS Russia). He continued his work with the Kiev MESM team in parallel. He was invited to the IPMCE by M.A. Lavrentev, than its director, who offered Lebedev the position of chief designer of a new computer called the BESM ("Bystrodeystvuyushchaya Electronnaya Stchetnaya Mashina" - Fast Electronic Calculating Machine). Soon after that MESM got the name of the Small Electronic Calculating Machine (in Russian – Malaya Electronnaya Stchetnaya Mashina): Again MESM, but no longer the "Model".

Lebedev brought his own project draft from Kiev, where he represented his view on the BESM architecture. One of the IPMCE best designers, Petr Golovistikov (shown in Figure 3), in [8] wrote the following:

> *"There exists a legend that Lebedev drew the whole BESM scheme on his "Kazbek" cigarette boxes or on separate paper sheets. That's wrong. His project description consisted of several thick notebooks. One could see there the detailed schemes and timing diagrams, including many different versions of different operations production."*

Fig. 3. Petr P. Golovistikov (the 1950s)

Fig. 4. Cover of Lebedev's working notebook

V.B. Karpova, collaborator of the IPMCE museum, identified among other IPMCE documents one of those notebooks (see Figure 4), which had been purchased in Kiev. The notebook consists of 100 grid sheets and is filled out in Lebedev's own hand. Among the titles that are in the notebook, we can see the following.

07.07.50:	External memory management (magnetic recording).
09.07.50:	Sending data from magnetic tape to the drum.
12.07.50:	To consider the model version, which includes common elements for instructions and numbers with one control switch that works on 4 clock pulses (not 3).
16.07.50:	Hardware and program controlled data transfer from tape to drum.
	Choosing number of bits for the machine word.
	Choosing the number of bits for the model with parallel number input.

21.07.50:   Translating from binary to decimal presentation using the computer.
            Operations.
23.07.50:   Magnetic tape control.
04.08.50:   Possibility and reasonability of parallel code input and trigger sells memory implementation.
08.08.50:   Sending data from the drum to internal memory.
12.08.50:   Multiplication with sequential code input.
            Flow-chart of executive unit.
            Flow-chart of dynamic memory on electro-acoustic tubes.
            Adjusting the IM (instruction management) from CCU (central control unit).
            Organizational of work.
            Deliverables.
            Finally.Calculating the remainder absolute value.
            Multiplication with 2n bits output.
            Addition with 2n bits.
            Developing the method of special operations production.
            Addition with exponent blocking.
            One-address instruction variant.

Reading of this notebook helps us to follow the entire process of gradual (though very fast) understanding of the future computer structure. The first date marked in the notebook is 7 July 1950, when Lebedev was already working on both projects – bringing the MESM model to operational mode and developing the first "real" computer BESM. The last date (12 August 1950) is denoted on page 46 (see Figure 5). The subsequent notes are not dated. This means that it took only one month to fill the half of the notebook. Obviously the second half was filled out by the end of the summer of 1950. During this period, Lebedev developed basic units of the future computer and the algorithms of performing (_producing_ as it is written in the notebook) of basic operations: addition, multiplication, division.

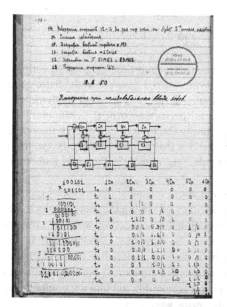

Fig. 5. Pages 46 and 47 of Lebedev's working notebook

It was necessary to calculate everything, even the length of magnetic tapes, intended to be external memory. The calculations like this may be seen on pages 6 and 7 (see Figure 6). As a result, the total tape length was estimated (approximately 200 meters) and the total time of data transfer from tape to drum (about 20 minutes). Moreover, here the remark is, *"The time is acceptable "*.

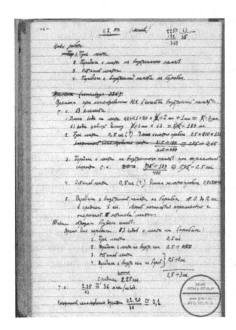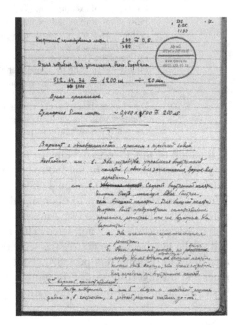

Fig. 6. Pages 6 and 7 of Lebedev's working notebook

Lebedev described all operations in details, at the same time he tried to estimate the time needed to perform those operations, trying to find answers to the basic questions of the starting period of every new computer design – is it worth implementing this or that operation, what would be the overall performance of the computer? For example, on page 47 (see Figure 5) after calculating the time of performing the operation of getting the value of $1/x$ (with precision not less than 2^{-30}) there is the result written: 1,5 milliseconds and the resolution reads: *"Acceptable"*. Pages 84 and 85 are of special interest (see Figure 7).

These two pages contain notes which are not about technical details (see Figures 8 and 9): they are focused on organizational principles of the big computer project. The notes were definitely made during the preparation for a meeting with government authorities, who were the key persons for the new computer development. In the notes, we can see the concerns with which both development teams (Kiev team and Moscow team) encountered. The number of problems is not small but the conclusion is quiet simple – with a real help for the project, it is possible to build the computer by the second half of 1952. It is surprising but the terms were met.

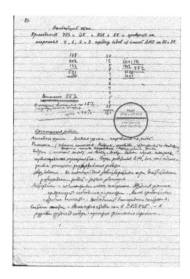

Fig. 7. Pages 84 and 85 of Lebedev's working notebook

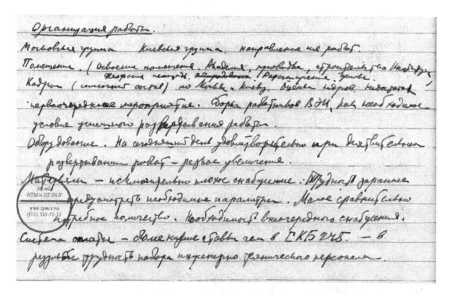

Fig. 8. Lebedev's working notebook fragment devoted to the organizational problems of two development teams

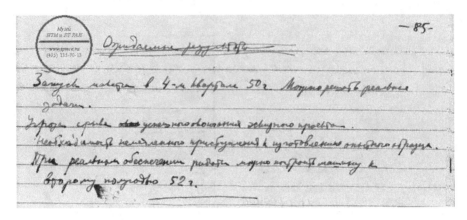

Fig. 9. Lebedev's working notebook fragment devoted to the BESM project expectations

3   Lebedev's Organizational Notes

V.B. Karpova made the following decryption of the organizational notes and project plans written by Lebedev.

Organization of work (page № 84 of Lebedev's notebook).
o   *Moscow team. Kiev team. Project directions.*
o   *Offices (preparing the offices. Academy, electrical wires, building the Institute*
o   *Theophania (location of the Kiev laboratory). Space. Increasing offices. Apartments)*
o   *Personnel (the list) in Moscow and Kiev. Personnel estimation. Need for more.*
o   *Immediate task. Gathering employees from Electrical Institute (VEI-All-Union Electrotechnical Institute) is the necessary condition of successful project development.*
o   *Equipment. For today satisfactory, for real development project – needs a sharp increase.*
o   *Materials – exceptionally poor supply. Difficult to foresee beforehand the needed parameters. Relatively small needed amount. Necessity in out of order supply.*
o   *Salary system – lower rates than in SCB 245 (Special Design Bureau No 245 of Moscow with its "Strela" computer – BESM competitor) – as a result, the difficulties with hiring of engineers and technical assistance personnel.*

Project expectations (page № 85 of Lebedev's notebook).
o   *Model launching in the 4th quarter of 1950. Possibility to solve real problems.*
o   *Hazard to wreck the successful completion of preliminary project draft.*
o   *The need of immediate start of building the experimental model.*

○   *Having the needed supply it is possible to build the machine by the second half of 1952.*

The list of note titles itself shows that Lebedev was doing his design work very carefully, trying to understand all the minutest details, many of which were discovered by the scientist for the first time. He also tried not to forget the importance of perfect project organization and planning. Lebedev himself wrote the BESM project plan, which was identified by V.B. Karpova. This plan was found among the documents in the IPMCE archive storage. Lebedev permanently controlled the progress (Figure 10). According to this plan, the preliminary project draft should have been finished by the end of the first quarter of 1951 (approximately the same time with the completion of MESM model). The plan designated three main work directions:

1.   Cells principle schemes design.
2.   Mathematical and experimental testing.
3.   Preliminary project draft development.

Fig. 10. BESM preliminary project draft plan handwritten by Lebedev in 1950

For each plan direction twenty basic items were defined, for which the dates and resources were fixed. Those basic works were the following:

1.   Development of the basic electronic elements (counters, cells of static storage, keys, accumulators).

2. Development of the methods of producing the arithmetic operations (addition, subtraction, multiplication, division).
3. Development of the methods of producing the special operations (comparison, shift, negation, interpolation).
4. Development of the methods and auxiliary devices for converting numbers from decimal to binary system and backward.
5. Development of the arithmetic unit.
6. Development of the unit for interpolation.
7. Development of the auxiliary units for solving the system of linear equations.
8. Development of the fast internal memory unit.
9. Development of the external memory unit.
10. Development of the central controlling unit of the machine.
11. Development of the instruction management unit.
12. Development of the operation management unit.
13. Development of the memory management unit.
14. Development of the unit for program and external digital data preparing.
15. Development of the unit for final result storing.
16. Development of the inter-unit link system.
17. Development of the monitoring, signaling and power supply system.
18. Development of the overall computer scheme.
19. Development of the computer design and construction.
20. Development of the computer working model (using the decreased frequency, decreased number of bits, restricted memory size, built on static electronic tubes and with restricted number of operations).

4   From MESM to BESM

The plan was written not for Lebedev alone. By the spring of 1951, nearly fifty engineers have been working in the IPMCE Laboratory No 1. There were experienced employees such as L.A. Lubovitch and K.S. Nesloukhovsky on his team, as well as very young people and future academicians such as V.S. Bourtsev and V.A. Melnikov. The future IPMCE chief designers V.N. Laut and P.P. Golovistikov (who at 27 years of age was four years older than others were) and some other very successful IPMCE collaborators were also among them. Young colleagues of Lebedev who simultaneously monitored two fundamental projects of MESM and BESM, were very helpful for him and his work.

The changes can be seen on the plan. Sometimes the initial dates were shifted, but some of them were later restored. Still the preliminary project draft was completed just during the first quarter of 1951. It was very important for Lebedev to run his both projects without hindering each other. In the very beginning of 1951, the MESM presentation to the state commission took place. On the 21st of Aprilof 1951 the state commission on BESM project draft started its work. By the end of December 1951 MESM operated in its new design and in summer of 1952, the BESM manufacturing was completed. It was in operation by the autumn of 1952.

The work on BESM was not an easy one. The Special Design Bureau 245 was working on its own computer called the "Strela" (meaning "Arrow") in parallel and in certain circumstances only one computer (BESM or "Strela") could be chosen for serial production. The traces of scientific and technical competition can be seen in Lebedev's notebook (see page 84 of Figure 8). As a result, in April 1953 the State Commission with its chairman M.V. Keldysh approved the successful completion of BESM development. Since June 1953 Lebedev had started his work as the IPMCE director. Now this institute is named after him.

On 23rd of October 1953 Lebedev was elected a member of the AS USSR within the Division of Physics and mathematics. He was the first academician with specialization "calculating devices". In 1954 he was rewarded by the USSR government highest award for creating the BESM computer and in 1956 he was given the USSR highest honorable title "Hero of Socialist Labour".

Fig. 11. BESM-1 computer on the ground floor of IPMCE building in Moscow

Many problems earlier seemed to be unsolvable, because of the large amount of calculations, were easily solved using BESM. Lebedev himself liked the example of calculating the artillery shell flight trajectory. With BESM, this calculation was faster than the shell itself.

Fig. 12. Serial version of the BESM computer that was named BESM-2

Fig. 13. O.K. Shcherbakov (to the right) and A.A. Pavlikov together with the member of Chinese group in IPMCE

The first BESM (Figure 11) was installed on the ground floor of the IPMCE building in Moscow. For the long time it was solving both scientific/theoretical and application problems (some of them were described in Lebedev's article in [3]). In particular, this computer was used for calculating the trajectory of the rocket that brought the USSR emblem to the Moon.

Fig. 14. V.A. Melnikov (to the right) during his business trip to China

After establishing of the Computing Centre at the AS USSR in 1955, the new task had to be performed by IPMCE: to prepare BESM computer for mass production. That was done by the end of 1957, when a factory in Ulyanovsk (a city on river Volga) started to manufacture the computer named BESM-2 (Figure 12). These computers were installed in almost all computing centers of the country. BESM-2 was used for calculating the data for the Earth satellites launching and for the first manned space ships.

BESM-2 was reproduced in China with the help of IPMCE team of engineers headed by Oleg Shcherbakov [2]. V.A. Melnikov also visited China. Five IPMCE collaborators were given state awards of the Chinese People's Republic. There are several pictures with IPMCE engineers and their Chinese colleagues in the IPMCE museum (Figures 13 and 14).

In October 1955, Lebedev made his report at the International Conference for electronic calculating machines in Darmstadt (West Germany). That was a real sensation: BESM appeared to be the fastest computer in Europe. The future would have shown that all the computers of the BESM series (from the first BESM to BESM-6) were the best universal computers in Europe for the moment of their first appearance.

Getting such powerful instrument as BESM for complicated mathematical calculations did not stop development of new computers. By that time, Lebedev could clearly see the important role of computers in the state scientific and economic progress. By the moment of BESM's completion Lebedev had already thought over the principles and architecture of a new computer – M-20 that should have become the fastest computer in the world (its performance was about 20000 instructions per second). Many new logical operations were implemented in M-20 such as address modifications, operation combinations, and hardware loop support.

Besides general applications Lebedev made a great contribution in another computer implementation field – controlling objects in a real time mode. For his colleagues and the future IPMCE director V.S. Bourtsev (Figure 15) among them, Lebedev formulated the task of building small computers "Diana-1" and "Diana-2", that were intended for using in the systems of directing a fighter to its targets[1].

Fig. 15. V.S. Bourtsev

Yesterday's students soon became famous scientists who carried on the development of new computers. Academician V.S. Bourtsev started his carrier in IPMCE with designing the BESM instruction management unit. He continued with taking part in the design of a series of "Diana" computers, which were followed by the M-40, M-50, 5E92b, 5E51, 5E26 and the "Elbrus" series. The first work of (future) academician V.A. Melnikov started with BESM operation management unit, later he continued it with BESM-2, BESM-6, and the AS-6.

In the IPMCE museum, there is a model of one of the BESM-1 sections (Figure 16), which is a real monument to science and engineering. This model represents original units, cells and separate fragments of BESM-1 and BESM-2 that remained intact since the beginning of 1950s.

The value of BESM computer for Russian and world computer engineering can hardly be overestimated. Many of that, what was first tested in process of this computer design, is an ordinary ("classical") thing now [4,6,7].

Fig. 16. Model of BESM-1 computer in the IPMCE history museum

References

[1] Bourtsev, V.S. (ed.): Sergey Lebedev. 100 years since birthday of the founder of domestic electronic computers. FizMatLit, Moscow (2002)

[2] Lebedev, S.A.: The creator of domestically produced computers, 2nd edn., Institute for Precise Mechanics and Computer Engineering (1990/2002)

[3] Lebedev, S.A.: Electronic Calculating Machine, Pravda newspaper, Moscow (December 4, 1955)

[4] Lebedev Institute of Precision Mechanics and Computer Engineering, Russian Academy of Sciences (IPMCE), http://www.ipmce.ru/

[5] Malinovsky, B.: Store Eternally. Gorobetz, Kiev (2007) ISBN 978-966-96940-0-3

[6] Museum of the USSR Computers History,
http://www.bashedu.ru/konkurs/tarhov/english/index_e.htm

[7] Russian Virtual Computer Museum,
http://www.computer-museum.ru/english/index.php

[8] Ryabov, G.G. (ed.): From BESM to supercomputer. Notes on history of Lebedev's IPMCE by his colleagues. Institute for Precise Mechanics and Computer Engineering, in 2 volumes (1988)

[9] Zhdanov, P.S., Lebedev, S.A.: Stability of parallel functioning of electrical systems, 2nd edn., pp. 263–387. EnergoIzdat, Moscow (1933/1934)

Some Hardware Aspects of the BESM-6 Design

V.I. Smirnov

Institute of Precise Mechanics and Computer Engineering (IPMCE)
Russian Academy of Sciences, Moscow, Russia
svi@ipmce.ru

Abstract. This paper very shortly describes some hardware solutions of central processor (CPU) of BESM-6. CPU had very deep instruction pipe with an associative buffer for instructions and an associative buffer for data with original protocol. Logical and storage elements used only domestic discrete components. Main logical unit based on differential amplifier with pyramid of rich diode logic and paraphase synchronization. Original construction without printed plate made wire connections very short and gave possibilities for direct access to every contacts and interchanging modules. All these solutions permitted to achieve high clock frequency, reliability and effective maintenance.

Keywords: General-purpose computer architecture, instruction pipe, logic circuit design, mechanical package.

1 Introduction

The Institute of Precise Mechanics and Computer Engineering (IPMCE) developed the general-purpose computer BESM-6 under the leadership of academician S.A. Lebedev. It became operational in 1967. The BESM-6 integrated the latest architecture features and domestic experiences with many original solutions that provided high efficiency and exclusively a long period of exploitation.

First computers were installed in major scientific and cosmic institutions and had played important role in research and development domestic science, defense and technology. Later BESM-6 had undergone several modifications and successfully operated in different fields of industry and education as one of the most popular general-purpose computer. Up to four hundred computers were produced with modifications of input/output and memory parts and some computers are known in working state nowadays.

This paper is devoted to some hardware aspects of the BESM-6 central processing unit piping not reflected in other paper about software.

2 BESM-6 Characteristics

Figure 1 shows the front view of the BESM-6 CPU in its working state in 2006. Some of its general characteristics include the following:

J. Impagliazzo and E. Proydakov (Eds.): SoRuCom 2006, IFIP AICT 357, pp. 20–25, 2011.

- o   48-bit word with 2 parity bits in memory (one for each half-word)
- o   Two 24-bit single-address instructions with two formats in one program word
- o   General-purpose operation set with floating-point arithmetic
- o   15-bit address space with extension possibilities
- o   Wide set address transformations with stack mode included
- o   Page mode protect system for memory
- o   Privilege mode program execution
- o   Extended concurrent i/o operations
- o   Hardware support for interrupts
- o   Efficiency and exclusive reliability, with other features

Fig. 1. Front view of the BESM-6 central processing unit

Of particular importance in achieving high performance was flexible instruction pipeline for the CPU with large throughput and deep look ahead. Its integration with all previously mentioned characteristics permitted to create large amount of mathematical software and application packages. During the BESM-6 lifecycle, rich software was designed including compilers for all known programming languages, operating systems, and application packages as the largest domestic soft fund.

3   Basic Architecture

BESM-6 pipeline was designed independently of well-known Stretch computer; its chief designer S.A. Lebedev called it a "water pipeline". It was investigated in detail on a program model. The pipeline architecture of the BESM-6 CPU actually had all features of contemporary microprocessors but with very limited resources.

Original logic elements with paraphase synchronization allowed creating flexible command pipeline for the CPU with 300 ns instruction cycle time and deep look ahead (six and more instructions). Unique torodial ALU with very fast cycle time. The general block-diagram of BESM-6 central processing unit appears in Fig. 2.

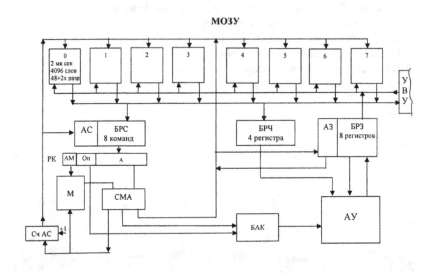

Fig.2. General block-diagram of BESM-6 central processing unit

Main functional units:

МОЗУ	–	magnetic ring storage with blocks of 4096 (48+2 bit) words with 2 microseconds cycle time.
АС-БРС	–	associative buffer (8 instructions) with 300 ns cycle time.
РК (АМ, Оп, А)	–	instruction register, where:
	АМ	– address of index registers (M),
	Оп	– operation code, and
	А	– operand address.
СМА	–	adder for address modification.
БАК	–	buffer for arithmetic commands.
БРЧ	–	buffer for operands from memory (300 ns).
АЗ-БРЗ	–	associative buffer for results written to the main memory (300 ns). The oldest unused result was transferred to memory.
АУ	–	toroidal multifunctional ALU with many interesting solutions for special paper.

The control unit coordinates control the operation of all functional units and it includes an interrupt system, I/O operation, and many other functional units to support effective work of operating system.

The base for the system of semiconductor logical elements was a differential amplifier with common emitter and wide pyramid of diode logic (Fig. 3). It included back link gates for the latch function and had an AND-OR-AND-pyramid to set the latch in definite state. One may say that this logical element base supported original construction solutions that allow the design of a very effective and reliable computer.

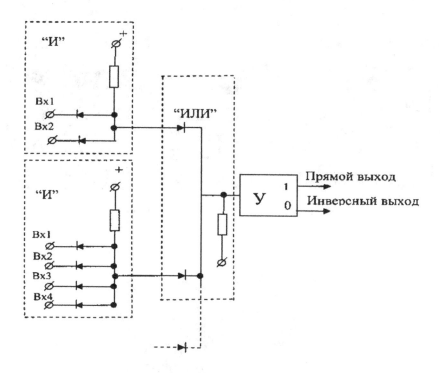

Fig. 3. Basic logical unit

Four amplifiers with backward gates were placed in one physical block (Fig. 4) with their state indicators. Input logic was in other physical block (Fig. 5), which was placed on the opposite side of chassis as close as possible to keep wires short (Fig. 6). It is the use of all these constructive solutions with an original paraphase synchronization based on two sinusoidal signals with 10 MHz frequency that permitted a clock of 100 ns in pipeline processing.

The integral use of all mentioned solutions allowed them to reach a rate of 1 MIPS using ordinary domestic semiconductor technology, which had the best components per operation ratio.

Four chasses with У-blocks were placed in one cabinet and four adjoining cabinets, showed in Fig. 1, display all CPU latches and permit direct physical access to all contact points for oscillograph.

The test and diagnostic system provided extremely high results in maintenance and finding accidental faults. All this made possible extremely long periods of reliable operation in many computing centers.

Fig. 4. Fragment of chassis with У blocks

Fig. 5. Д block

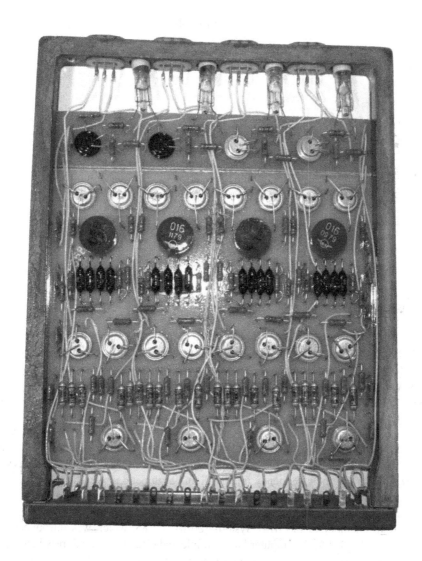

Fig. 6. Y-block

Computer Architecture Development: From the BESM-6 Computer to Supercomputers

Yuri I. Mitropolski

Institute of System Analysis
Russian Academy of Sciences
mitr@imvs.ru

Abstract. This work outlines the architecture and main features of the BESM-6 computer and data processing system AS-6. It also discusses the development of the "Electronika SS BIS-1" supercomputer system as well as the principle results of the research on multi-architecture (heterogeneous) supercomputer systems.

Keywords: Supercomputer architecture, heterogeneous multicomputer systems, pipelined vector processors, multi-architecture supercomputer systems, scalable heterogeneous chip multiprocessors.

1 Introduction

The BESM-6 computer became the milestone of domestic computer development in the USSR. The chief designer of BESM-6 was academician S.A. Lebedev; the vice-chief designer was academician V.A. Melnikov. This computer was oriented towards the solution of large-scale scientific and engineering problems. This fact was the major influence on the architecture, the logical design, and the construction of the machine.

The logical circuits were based on diode gates and transistor current switch amplifiers with a floating power supply. Due to the original construction with short wires between modules, it achieved a very high clock frequency of 10 MHertz. It used a pipeline on clock frequency in control and arithmetic units. The interleaving of banks of the main memory and non-addressed buffer memory with associative search maintained the proper speed of access to the memory. In fact, this design was similar to the cache memory of the IBM System 360 Model 85, which was introduced two years later. The BESM-6 computer had a simple and effective one-address instruction set, a floating-point arithmetic unit and registers for address modification. For multi-programming, it used paged virtual addressing, a processor interruption system, a privileged instruction set, and a multi-user operational system. There were seven selector channels for magnetic drum, disk and tape memory, and multiplexer channel for input-output devices based on CPU multiprogramming. The original design methodology had two features: a formula description of logics and a table system with a table for each module describing all its wire connections [4].

J. Impagliazzo and E. Proydakov (Eds.): SoRuCom 2006, IFIP AICT 357, pp. 26–30, 2011.

2 Developments

The beginning of the production was 1967. The BESM-6 computer set two records: the duration of production of fifteen years and the duration of use for more than twenty years. The impact of this computer on many generations of engineers and programmers was based mainly on the fruitful ideas of its original design.

The next major development produced under the leadership of chief designer V.A. Melnikov was the AS-6 data processing system. The installation and maintenance of the BESM-6 at the computer centers that processed large volumes of data from many abonents, in particular from the Space Flight Control Center, served as a stimulus for the AS-6 system development. A weak point in the BESM-6 design was the small number of external devices available for use and the low throughput of the input-output channels. At the first stage the main task was the design of the interface unit (in Russian Аппаратура Сопряжения - АС-6) between the BESM-6 machine and AS-6 machine with the capability of attaching a large number of telephone and telegraph channels, channels for telemetric data, and also increasing the number of peripheral devices. However, after experimental work with at the first stage it became clear that the system must contain more powerful means of processing data; most of all, it must be a scalable system with the ability of attaching additional computers and devices.   In the end, the task was the design of multicomputer system with an advanced means for system reconfiguration.

3 The AS-6 System

The fundamental ideas were the creation of application-specific subsystems and devices, and a unification of channels for the whole system. Beside the BESM-6 the system included an AS-6 central processing unit, a peripheral computer PM-6 consisting of a peripheral multichannel processor, a multichannel exchange unit (channeler), additional units of main memory, magnetic disks controllers, and a telemetric controller. They connected all these components by a unified first-level channel.

Since 1973, the AS-6 system was in use for experimental purposes.   At the same time its development was continued. In 1975, the system was in use during works on the joint Soviet-American "Apollo–Soyuz" mission. The complete system was ready in 1979.

In the AS-6 system design, they realized for the first time new ideas that became the basis for supercomputer development and research on advanced high-performance computer architecture. To begin, it is necessary to highlight the following features:

o The AS-6 was a heterogeneous multicomputer system;
o The AS-6 central processing unit was problem-orientated for the tasks of complex objects control and effective translation;
o The PM-6 peripheral computer was function-specific for input-output control;
o The system used several specially designed unifying channels.

4   The BESM-10 and the Electronika

During the period of development and operation of the system, it became obvious that new architectural ideas and old electronic technology were not corresponding one to the other. In 1973, they worked out the BESM-10 system. They planned to develop the advanced computer system on high-speed ECL LSI chips. The Ministry of Radio Industry of the USSR did not support this project.

Further development in this direction under the leadership of academician V.A. Melnikov was the creation of the "Electronika SS BIS-1" supercomputer system. The aim of this development was to increase the performance of the new computer by two orders of magnitude. This led to the development of a new technology and appropriate architecture [5].

It was planned on the first stage of design that the system would include the following problem-oriented subsystems: a main computer with a pipelined vector processor, a matrix computer, and a logical data processing machine. In addition to this, they planned the development of the following function-specific subsystems: a peripheral computer, an external solid-state memory controller, a magnetic disk controller, control computers, and front-end computers. Due to the shortage of resources, they decided to postpone the development of the matrix, logical and peripheral machines.

The two choices of architecture for the main computer came under review; namely, the AS-6 central processor as enhanced in BESM-10 project and the pipelined vector processor. The second one was more appropriate due to much higher performance and better utilization of synchronous pipelines. The feature of external solid-state memory was the development of intellectual controller intended for the realization of different access methods for the main computer and for joining up two vector machines in a single platform. The peak performance of two-processor system achieved 500 MFLOPS. The software included operational systems of main and front-end computers, programming systems for macro assembler, and the Fortran 77, Pascal, and C programming languages [6].

In 1991, they tested the "Electronika SS BIS-1" system. Four sets of system hardware were delivered to customers. In the same year, they developed the "Electronika SS BIS-2" multiprocessor system. The planned peak performance was 10 GFLOPS. Beside multiprocessor main machines, the system has included monitor machines for task preparation and system control as well as massively parallel subsystems. However, in 1993, they made the decision to discontinue all work on this effort.

5   Reflections

The experience gained during the development of these systems became the basis for the research on heterogeneous supercomputer systems. They showed that the most effective system was one with subsystems oriented toward different forms of parallelism, that is, those corresponding to different forms of parallelism of parts belonging to the very large problems. The processing of large data sets is connected with data level parallelism. This form of parallelism is used most effectively in vector processors. Another form of parallelism is parallelism of tasks, which is present in problems with great number of independent or weakly coupled subtasks. In this case, the best choice is a multiprocessing subsystem.

At the first stage of research, they developed the concept of heterogeneous supercomputer systems. In particular, they proposed to integrate in a single platform a vector uniprocessor and a multiprocessor, both having the access to the common memory [7]. The whole system could have several such integrated platforms based on large system memory [1].

The next stage of research concerned the analysis of advanced ULSI technologies with more than one billion transistors on a chip. They developed the original architecture of a scalable multi-pipelined vector processor. This processor had many parallel pipelined branches each having a chain of simple vector processor with several functional units and a local memory. A program having complex vector operations achieved the best results. The performance of such scalable processor was up to 1024 floating-point operations per cycle [8].

In accordance with these concepts, the research project of a multi-architecture supercomputer system had been developed [9]. The system consisted of a computing subsystem, monitor-simulating subsystem, system memory and external subsystem. The computing subsystem includes a vector multiprocessor, a scalar multiprocessor, and a monitor computer. The productivity of this system can be very high, up to 90% of its peak performance. The performance depends on the level of technology and it could reach up to 10 PFLOPS.

6 Conclusion

The comparison of this project with foreign researches and developments shows that in theoretical and conceptual sense it is ahead of published results [10]. On March 20 of 2006, Cray Inc. announced plans to develop supercomputers that will take the concept of heterogeneous computing to an entirely new level by integrating a range of processing technologies into a single platform. The second phase will result in a fully integrated multi-architecture system [3]. The same concepts were published in 1995 [7]. The term "multi-architecture" was proposed in 2003 [9].

The design of the cell chip by Sony, IBM, and Toshiba began in 2001. Cell is a heterogeneous chip multiprocessor that consists of an IBM 64-bit Power Architecture™ core, augmented with eight specialized co-processors based on a novel single-instruction multiple data [2]. In a paper published in 1998 [8], one reads the following:

> "The chains and parallel branches of the modules provide high performance due to their topology with the shortest links between functional units. In a maximum configuration, the uniprocessor might have up to 1024 modules and attain up to 4 thousands floating-point operations per cycle or 4 TFLOPS".

References

1. Anohin, A.V., Lengnik, L.M., Mitropolski, Y.I., Puchkov, I.I.: The architecture of Heterogeneous Supercomputer System. In: Proc. Fifth International Seminar, The Distributed Information Processing, Novosibirsk, pp. 22–27 (1995) (In Russian)
2. The Cell Architecture,
   http://domino.research.ibm.com/comm/research.nsf/pages/r.arch.innovation.html
3. Cray Will Leverage an "Adaptive Supercomputing" Strategy to Deliver the Next Major Productivity Breakthrough, http://www.cray.com
4. Kuzmichev, D.A., Melnikov, V.A., Mitropolski, Y.I., Smirnov, V.I.: The Principles of Design of Large-scale Computers. In: Proc. Jubilee Scientific Conference Devoted to 25 Years of Institute for Precise Mechanics and Computer Engineering USSR Academy of Sciences, Moscow, pp. 3–16 (1975) (In Russian)
5. Melnikov, V.A., Mitropolski, Y.I., Malinin, A.I., Romankov, V.M.: Requirements for Hardware Design of High Performance Computers and Problems of Their Development. In: Melnikov, V.A., Mitropolski, Y.I. (eds.) Proc. Cybernetics Problems, Complex Design of Electronics and Mechanical Construction of Supercomputers, pp. 3–10. Scientific Counsel on Cybernetics of the USSR Academy of Sciences, Moscow (1988) (In Russian)
6. Melnikov, V.A., Mitropolski, Y.I., Reznikov, G.V.: Designing the Electronica SS BIS Supercomputer. IEEE Trasactions on Components, Packaging, and Manufacturing Technology, Part A 19, # 2, 151–156 (1996)
7. Mitropolski, Y.I.: The Concept of Heterogeneous Supercomputer Systems Development. In: Proc. Fifth International Seminar "The Distributed Information Processing", Novosibirsk, pp. 42–46 (1995) (In Russian)
8. Mitropolski, Y.I.: Architecture of Multipipe-lined Module Scalable Uniprocessor. In: Proc. Sixth International Seminar "The Distributed Information Processing, Novosibirsk, pp. 30–34 (1998) (In Russian)
9. Mitropolski, Y.I.: Multiarchitecture Supercomputer System. In: Proc. First All-Russian Scientific Conf. "Methods and Means of Information Processing", pp. 131–136. Moscow State University, Moscow (2003) (In Russian)
10. Mitropolski, Y.I.: Mutiarchitecture – New Paradigm for Supercomputers. Electronics: Science, Technology, Busines, # 3, 42–47 (2005) (In Russian)

Operating System of the Multi-machine Computer AS-6

I.B. Bourdonov, V.P. Ivannikov, A.S. Kossatchev,
S.D. Kuznetsov, and A.N. Tomilin

Institute for System Programming of RAS
{igor,ivan,kos,kuzloc}@ispras.ru
tom11@bk.ru

Abstract. The S.A. Lebedev Institute developed operating system for the AS-6 distributed computing system (OS AS-6) in the 1970s. The OS AS-6 consisted of peer operating systems of separate machines making up the AS-6 computer system. Those operating systems interacted through a uniform interface. The OS AS-6 facilitated interaction between computation processes on the nodes of the AS-6 system through a network together with their interaction with the global network. It provided the possibility for any process in the AS-6 system to use all the devices connected with the nodes of the AS-6 computer both with addressing and with usage of external devices. It also supported a pipeline operation of the computers in the AS-6 system performing real-time processing of large data streams of spacecraft missions. The operating systems of the nodes of the AS-6 system had a special means for parallel processes management, including task hierarchy organization and information processing management within a single task.

Keywords: Distributed computing system, operating system.

1 Overview

In the 1970s, the S.A. Lebedev Institute of Fine Mechanics and Computer Engineering developed the operating system (OS AS-6) for the AS-6 distributed computing system [1]. The OS AS-6 consisted of peer operating systems of separate machines making up the AS-6 computer system. Those operating systems interacted through a uniform interface.

The OS AS-6 facilitated interaction between computation processes on the nodes of the AS-6 system through a network together with their interaction with the global network. It provided the possibility for any process in the AS-6 system to use all the devices connected with the nodes of the AS-6 computer both with addressing and with usage of external devices. It also supported a pipeline operation of the computers in the AS-6 system ("computer pipeline") performing real-time processing of large data streams of spacecraft missions. The operating systems of the nodes of the AS-6 system had a special means for parallel processes management, including task hierarchy organization and information processing management within a single task.

J. Impagliazzo and E. Proydakov (Eds.): SoRuCom 2006, IFIP AICT 357, pp. 31–35, 2011.

There were two kinds of network software in OS AS-6: transport and functional. The purpose of the transport software was to transport data between operating systems of separate nodes and between different user tasks in the system. The functional software evolved from the transport software and it carried out a variety of operations such as resource request processing, data exchange with input-output devices, and invocation of middleware programs. Since the operating systems of all the nodes were peers, their transport software (in contrast with the functional one) was identical. If some computer within the AS-6 system failed, the network would interact and stopped it; it automatically resumed after restarting its operating system. We used transport software typically for interaction between processes on different computers, except when a special computer managed external devices ("peripheral machines"). For the operating system of a peripheral machine, the transport software also supported interaction between internal processes.

2   AS-6 Mail Functionalities

Interaction between processes in a distributed real-time system should be very effective and have minimum overhead. Because of this, we organized the data transfer mechanism between peer operating systems based on high-speed channels and two-level transport software. The first level, called the *physical mail*, was responsible for message transfer through the high-speed physical channels. The second level, called *user mail*, was responsible for message transfer between user tasks and system processes. The physical mail effectively handled hardware faults such as processing unit outage, incorrect transfer of message data (detected by monitoring circuits), and message loss in the physical channel. The physical mail service of one unit in AS-6 could interact with fifteen parties, which were the physical mail services of the other units.

The main elements of a physical mail service were the following.

o   *Physical port*: These are the sending and receiving ports; when connected, these ports make up the data transfer lines.
o   *Messages with responses*: A sending port can send a message to a receiving port and the latter can send a response to the former. Each message had a response mechanism.
o   *Physical mail request*: Such a request contained a command of the physical mail service and its parameters, including the address of the message beginning and the message's length. Each request contained a reference to the port that should process this request.
o   *Directives*: A directive is an interrupt word by which one unit in AS-6 system can send it to another unit in the system.

The physical mail service used two kinds of directives: direct and inverse. A direct directive contained an address of some request. An inverse directive contained a response to some direct directive. A positive response occurred if the request processed normally; otherwise, it was a negative response. The transfer commands used by a physical mail request included, create or destroy the communication line, send a message and wait for response, receive a message, and send a response.

3 Communication Elements

Both sending and receiving ports could perform the creation of a communication line. Each port in AS-6 system had the unique number. Default or static mail subscribers such as resource managers had standard ports that statically defined numbers. Static subscribers could call occasional subscribers such as device control programs. At any moment of a call, they received ports with numbers generated on the fly.

A command to create a communication line for a receiving port did not cause any data transfer; it only created a record in the rendezvous table of the physical mail service. The service then waited for the create command from a paired sending port. A command to create a communication line for a sending port generated a direct directive referring to this request. When the receiving unit obtained such a directive, its physical mail service looked through the rendezvous table for a request to create a communication with the corresponding sending port. If it found such a request, the inverse directive contained a positive response along with the address of the receiving port; otherwise, it contained a negative response.

After creating a communication, the receiving port could process the commands to receive a message and to send a response. The sending port could send a message while waiting for a response. A command to receive a message created the corresponding direct directive to the sending port. After receiving this directive, the sending port would send a message.

A command to send a message generated a corresponding direct directive to the receiving party, if it had already sent the directive of the kind described above. The directive to receive a message forced the receiving party to read the message data and to synchronize with the subscriber, which then issued a command to receive a message. The processing of a command to send a message ended only after receiving a response to the message. A command to send a response generated a corresponding direct directive. A response could come asynchronously; that is, responses for two messages sent in some order could arrive in reverse order. To match the responses with their corresponding messages, each response contained the address of the request containing its message.

A special initial message sent to another party marked the initial availability of a physical mail service. To monitor the availability of each other, the physical mail service would send each party a special availability message. If for a definite time the service did not send availability message to its peers, it destroyed all the communication lines. The same thing occurred if other parties obtained an initial message from the service. This signaled a quick restart of the corresponding node due to the absence of the availability messages.

The user mail service helped to exchange messages for the user or system tasks when performed on any node of the AS-6 system. The interacting tasks by user mail could occur on one node, on different directly connected nodes, or on nodes having no direct connections. Commands performed by user mail service did not depend on the location of the interacting task, so the user mail service was a universal task interaction medium.

Subscribers were tasks interacting through user mail service. Each active (online) subscriber had an identifier unique for the entire network. Subscribers could be standard or temporary. A standard subscriber had one permanent identifier; a temporary identifier could have different identifiers on different activation periods. An identifier used by a temporary subscriber could transfer to another temporary subscriber if the first one became inactive. Some examples of standard subscribers include subscriber activation tasks, file transfer tasks, and network output buffering tasks.

A message exchange between two user mail subscribers occurred through a virtual communication line called an association. One subscriber could participate in many associations and in general, there was no limit to the number of associations between two subscribers. A port defined the point of interaction between a subscriber and an association. Two connected ports define an association, which always provided a full-duplex communication.

Data transfer between different tasks could be asynchronous or session-based. The first kind of interaction, called "mail-like", created an association between interacting tasks only at the moment of data transfer. Interaction of the second kind, called "phone-like", preserved this association throughout the session. Mail-like messages transferred independently on each other and in general, they did not preserve their order. A subscriber could receive messages in a different order than the order of their sending. Some mail-like messages could require approval of their receipt.

In the phone-like mode, before the transfer of any data, the system would commutate the association. Special commands closed a communication session. During a session, the user mail service could perform the following tasks.

o   Automatic receipt approval of all messages,
o   Automatic support of message order preservation,
o   Message transfer only if the input buffers of the receiving party have the
     appropriate empty space,
o   Interrupt transfer.

Messages were not sufficient to perform all practical tasks during phone-like communication. Interrupts were used for fast transfer of data of bounded volume (with a speed higher than the speed of a message) and to remove some messages from the association. The system would use the mail-like mode for occasional interactions and used the phone-like mode for intensive interaction on a considerable time interval.

4 System Devices

The operating system of the AS-6 computer directly connected with the corresponding device and for a device, it could assign tasks on different computers. It had special modules called resource administrators that could interact with each other with the help of the transport software (physical mail). Specialized modules of two operating systems supported the use of a device; one module was on the node where the client task was working and the other module was on the node connected to the device used.

A user would have the possibility to address the devices in the network. This occurred in two ways—by the administrative division where the device was located or by the direct device address. One would use the first mode, for example, to send an output of some task to a device close to the one from which it had taken an input, preferably located in the same room. It was possible to connect different devices from one administrative division with different computers. However, different administrative divisions had no device in their intersection; thus, any device could contain only one administrative division.

5 Conclusion

In this brief paper, we have described only few, very basic features of the OS AS-6. Here we would like to highlight some other aspects that are important in the historical perspective.

At first, as we know the OS AS-6 was the first heterogeneous distributed operating system. Their component OSs based on different internal architectural principals and worked together based only on uniform protocols and interfaces. Moreover, these component OSs were relatively autonomous, and some of them provided services even without any support of other ones.

Second, the above architectural principles were very interesting in their own right. For instance, in the design of some of these OSs we actively used principles that later were called *object-orientation.*

And last but not the least: all this work was done with very close cooperation and collaboration with our customers and friends from Space Control Centers. These joint activities allowed to use OSs and applications to support the unique Soviet-American space mission Apollo-Soyuz and many other critically important projects.

Reference

1. AS-6 Data Processing System. Russian Virtual Computer Museum,
   http://www.computer-museum.ru/histussr/as6.htm (in Russian)

Distributed Systems for Data Handling

Vladislav P. Shirikov

Joint Institute for Nuclear Research
Dubna, Moscow, Russia
shirikov@jinr.ru

Abstract. This article provides a brief historical review in the field of creation and use of distributed systems for computerized accumulation and information processing (handling) in Soviet Union, and especially in its scientific centers such as Joint Institute for Nuclear Research and not only. The author tries to demonstrate- how the chosen methods and ideas permitted to realize the systems, which could be named as "PreGrid-complexes" in Russia.

Keywords: Data handling, distributed computing, GRID.

1 Introduction

The creation and use of distributed systems for computerized accumulation, transmissions, and information processing in interests of scientific research started less than fifty years ago. Apparently, the first such system in Soviet Union occurred in 1961 for radio-channel transmission of Atlantic Ocean objects information to first Soviet semiconductor universal computer DNEPR in the Institute "Cybernetics" (Ukrainian Academy of Science) in Kiev. Basic realizations of multicomputer complexes for distributed jobs and data handling started in 1960s and the 1970s practically in all prominent research organizations in Soviet Union and abroad.

At the "top of success" of their development, we would have Grid systems in the nearest future; that is, complexes that correspond to three criteria according to the definition of one of Grid technology's ideologists, Ian Foster. He states [1]:
"The Grid is a system that:

- coordinates resources that are not subject to centralized control …
- using standard, open, general-purpose protocols and interfaces …
- to deliver nontrivial qualities of service, relating for example, to response time, throughput, availability, security and/or co-allocation of multiple resource types to meet complex user demands, so that the utility of the combined system is significantly greater than that of the sum of its parts."

As an ideal, such a Grid system permits the ordinary user to input his/her job for batch or interactive processing without the necessity to define the job's "passport" (JDL-file) via attribute requirements. This includes the address of certain computer computing elements (CEs) as part of computing Grid resources. This is true if the system`s middleware has high-level services such as resource brokers (RB) and many of CEs, as it is now in large-scaled realized projects such as Enabling Grids for

J. Impagliazzo and E. Proydakov (Eds.): SoRuCom 2006, IFIP AICT 357, pp. 36–45, 2011.

E-sciencE (EGEE). For example, see http://lcg.web.cern.ch/LCG/ and http://egee-jra1.web.cern.ch; however, there is no such service in the American Open Science Grid (OSG) project. We talk about distributed systems of twentieth century, but it is necessary to say that many ideas realized in contemporary Grid structured were not quite unknown to specialists, which created systems more than thirty years ago, as it will be shown below. In other words, it is possible to say, that the Grid is the realization of "old" ideas based on new technologies.

2 Early Activities at JINR

At such a physical scientific centre in the Russian territory, as the Joint Institute for Nuclear Research (JINR), which was organized in 1956 and closely collaborated with Western physical centers, especially with European Centre for Nuclear Research (CERN, Geneva), problems arose. These situations demanded a common approach for applications realization and matched software environment in computing servers; this would make easier for the data and the programs to exchange at least with use of magnetic tapes in time of direct computer links absence. As to main type of processed information, then at JINR, CERN and other centers (e.g. Brookhaven, Berkeley, and Daresbury), this was a large amount of experimental data from accelerators detectors. At JINR, which also carried out theoretical and experimental research in fields of low and intermediate energy physics, it was necessary, in particular, to register and process data taken from spectrometers in experiments with the use of a fast neutron pulsed reactor.

According to these application problems, one of main tasks of the 1960s and later was the technical and software support of a whole sequence of measuring, preliminary and final data processing for different fields of experimental physics and also for theoretical physics. The chosen method to resolve this task was essential. It was necessary to start building distributed multileveled computing complexes with the use of perspective and standard measuring equipment and standardized programming tools for applications (such as compilers for high-level languages as FORTRAN. The tools had to be compatible with Western versions of program libraries for universal and special destinations including programming tools for algorithms parallelization and tools for user access to computing resources of different types. Main tasks at JINR, mentioned below, were initiated and guided (through July of 1989) by a corresponding member of the Academy of Science (USSR) N.N. Govorun, the chief of mathematical division in the JINR Computing Centre (1963-1966), vice-director (1966-1988) and director (1988-1989) of Laboratory of Computing Technique and Automation (LCTA).

The first stage of distributed systems creation started at JINR in 1962, when the simple structure of two connected by cable computers M-20 and KIEV in JINR Computing Centre could collect (via cable also) and process information from spectrum analyzer, located in measuring centre of Neutron Physics Laboratory shown in Figure 1. They also processed the punched paper tapes with information, prepared with use of special measuring devices, where the events, which happened in accelerator detectors and automatically photographed (in High Energy Physics Laboratory), were preliminary processed by operators. In 1965, this structure was modernized; they mounted an additional M-20 and they replaced the KIEV, which was not reliable, by the MINSK-22. They also used this structure for batch processing jobs in other fields of research. Its schema looked like the graphical presentation shown in Figure 2.

1962-1965. The first initial steps, when at least two connected computers became responsible for experimental data handling and other calculations

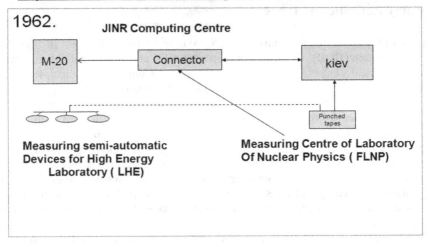

Fig. 1. Initial version of the JINR Grid

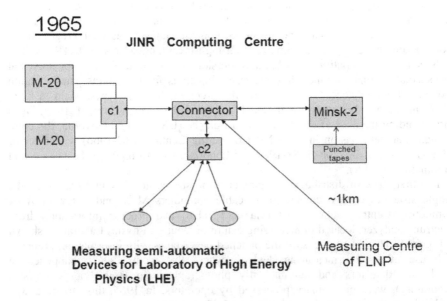

Fig. 2. Improved version of the JINR Grid

3 Late 1960s and Early 1970s Activities

From 1966 to 1974 saw the realization of the first stage of the three-leveled complex according to general principles and conditions (in addition to typical modern, up-to-date distributed systems for physics). The principles provide:

1. Meeting the requirements of online experiments;
2. Remote batch and interactive processing for jobs;
3. Interactive mode for users and their terminals;
4. Algorithm parallelism with use of different computing elements in complex;
5. Maintenance of archives for programs and data.

The realization of this set of services was defined at JINR by the appearance of a sizeable set of new computers. These included:

o "small" computers (first level) for on-line experiments (equipment control and data registration) or remote input/output stations (RIOS), used for batch processing support;
o "middle" computers (second level) for preliminary processing of data, taken from first level, and also for RIOS regime support; and
o "large" computers (third level) as basic computing servers.

This new set included some TPA (Hungarian production) computers, SM-3/4, M-6000 (analog HP-2116B), original HP-2116Bs, BESM-3M and BESM-4, MINSK-2/22, CDC-1604A, and the BESM-6. The BESM-6 started to be used as a basic computing server not only at JINR, but also in all noted scientific Soviet organizations. In fact, it was necessary to develop a preliminary software system. This is the reason the" DUBNA" appeared as a multi-language system with FORTRAN and other compilers and a programming library that was compatible with CERN library [2,3,4].

Another problem concerned the necessity to develop technical and corresponding software tools in BESM-6 operating system (OS) for connectivity with various local external devices and other computers of first and second levels. As a result, at JINR they modernized the BESM-6's peripheral control unit and new version of the operating system started to serve eight fast cabled lines. The lines had speeds up to 500 Kbyte/sec per line and up to 4 abonents connected to each line. In particular, they deployed remote data processing computers and RIOSes (also called "FORTRAN Stations" or FS) equipped with their own peripheral devices such as magnetic tape units, card readers, printers and terminals.

As far as the "middle" computers in laboratories are concerned, measuring centers had to participate interactively for information processing together with the BESM-6 with special "algorithm parallelism" tools. This is the reason to include a special "extracode" facility to serve external links embedded in the OS of the BESM-6. Additionally, they added subroutines to the DUBNA monitoring system that used this extracode useful in calling FORTRAN-written programs. Use of external interruption apparatus in the BESM-6 and peripheral computers (such as the BESM-4 that also used an analog of the FORTRAN-oriented monitoring system) helped to realize an analog of a modern Message Passing Interface (MPI) service. The schema of the complex in 1974 appears in Figure 3, where the CDC-6200 was added in 1972 as a BESM-6 partner on level 3 of the whole complex.

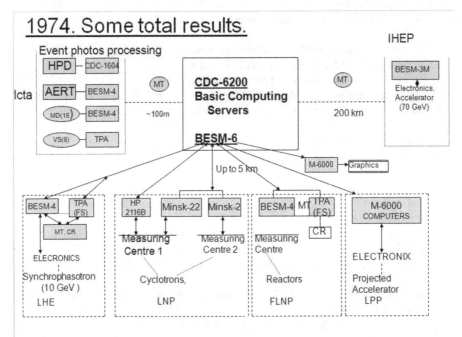

Fig. 3. Use of "middle" computers in the 1970s

The JINR was not unique in Russia, where they already had created multicomputer complexes in the 1960s. It is necessary to mention other systems in use. These include the AIST in the Computing Centre (Siberia), the complexes at the Institute of Applied Mathematics in the Russian Academy of Sciences (IAM/RAS), the Computing Centre at the Institute of Theoretical and Experimental Physics (ITEP/RAS), the Institute of Precise Mechanics and Computing Techniques (IPMaCT) in Moscow; and the Institute of High Energy Physics (IHEP) in Protvino.

In principle, the complex systems at JINR and the aforementioned systems had some similar features. For example, at CERN in system FOCUS, there existed concentrator (CDC-3200) for lines to servers CDC-6500/6400 (four cabled lines, up to 1Mbyte/sec and four peripheral computing or measuring abonents per line), RIOS stations, a terminal concentrator, and other peripherals. However, life was easier for Western scientific centers because there was no need for them to create for themselves the software environment as mentioned above. Regarding the type of computing servers, physicists in West always preferred to use 60-bit processor techniques for high-precise complicated calculations. This is why they used IBM computers for data registration and CDC serial 6000/7000 for processing and other calculations. Such computers were under embargo in the 1960s, but even the 48-bit BESM-6 was good enough.

It is necessary to mention that Grid started with the use of 32-bit technique in clusters, but the appearance of the 64-bit AMD and Itanium 2 processors immediately initiated the use of new PC-clusters in scientific centers. Hence, already by 2003, for the CERN "Openlab" project [5], several major physics software packages have been successfully ported on the 64-bit environment. For example, the Parallel ROOT Facility (PROOF) is a version of the popular CERN developed software ROOT for interactive analysis of large data files. In Russia, approximately at the same time, the Institute of Computing Mathematics in Moscow started to use a similar cluster with Myrinet 2000 bus for simulations, which earlier were based on use of access to the RAS Supercomputer Centre.

4   Late 1970s and Early 1980s Activities

In 1975 to the beginning of the 1980s at JINR, saw the installation of new ES-seria servers that were compatible with IBM-360/370; this led to the development of tools for terminal access to level 3 in computer complexes as shown in Figure 4.

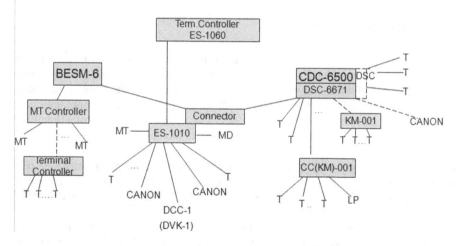

Fig. 4. Complex expansion using ES-servers

This development included creation of special controllers (multiplexers) for "large" ES-computers with the use of microprocessors. At JINR, they also developed the terminal concentrator based on the ES-1010 and a programming subsystem called INTERCOM. The system had the following features:

o Unified language (INTERCOM), which was originally used only for terminal access to CDC seria 6000) and tools for access to different computing servers via terminal concentrator ES-1010, equipped by its own external memory on disks and tapes and by a special operating system developed in JINR Laboratory of Computing Technique and Automation, former JINR Computing Centre. Alternatively, it used the ES-1060/1061/1066 microprogramed multiplexers. These ES-servers started to understand INTERCOM language by means of a process developed in LCTA software subsystem TERM.

o Independent (of BESM-6, CDC-6000) service for terminals in the process of information receiving, editing, and archiving, when users prepared their job texts.

o Execution of user directives (commands) to send a prepared job to a certain server (after automatically using the JDL-file edition) to show the status of jobs queues, to stop temporarily or kill a job in the server, or to print an output file on printer or terminal screen.

In principle, the concentrator could, in essence, execute functions, similar to functions of program services UI, NS and Resource Broker (RB) in modern Grid systems, especially if the user used a FORTRAN subset understood by all computing servers such as the BESM-6, the CDC, or the ES. The complex had the possibility to choose a server (this or that) simplified by the existence of compatible versions of library CERNLIB in the server's software.

5   Activities in the 1980s

The period 1983 to 1985 marked the starting point for the introduction of a standard means for the creation of local networks in scientific organizations and their external channels. Regarding local network structure and technology, the possible choices were the Token Ring/Token Passing or the then "young" Ethernet technologies. At JINR, they devised a local network in 1985. This was the JINET (Joint Institute NETwork) with a marker (token) method of access to a communication line of 12 km of 75-Ohm coaxial cable [4]. The standard adapter nodes were connected to it in various JINR buildings; each node served eight to twelve serial (standard RS-232C) ports for terminals or computer multiplexer ports. About thirty adapters were bought from the Swiss firm Furrer-Gloor; they had built-in Z-80 microprocessors and built-in memories.

Afterwards, the engineers at JINR (our LCTA engineers) produced a set of their modernized analogs and they used them in the frames of JINET. The LCTA programmers at JINR completely produced the supporting software for all adapters, because the firm's software was too expensive. Moreover, we foresaw the possible modernizations and wished to have not only built-in binary software, but also source texts. This software supported virtual communication channels (communication sessions) according to a user's application through a common cable with a throughput of about 1 Mbit/sec.

Each abonent (terminal user or computer) could give various commands to his connection node such as calling the text editor in his node, checking status of different nodes and their ports, setting the suitable transmission speed at his port or defining the terminating symbols for messages (end of text line, of package if the defaults were inadequate. If no specification was necessary, the abonent simply commanded to establish a virtual channel for communication with the given address (e.g. node and port number or the other computer's symbolic name), started to use this channel and disconnected at the end of session. Abonents were like computers and they could exchange files using the standard KERMIT program. A special node-administrator or his "hot reserve" partner automatically controlled the marker's (token) movements, collected statistics of nodes connections and faults, recreated the lost markers, disconnected every other node in case of noticed frequent malfunctions.

In 1988, JINET became the abonent of international X.25 standard IASNET, being connected to its communication center in Moscow [4]. This permitted to exchange files directly with physical centers, to use the telnet regime for access to certain servers; many JINR physicists had passwords for such access since they were members of different international collaborations. By the end of 1989, the standard JINR Ethernet was devised and connected to JINET as shown in Figure 5.

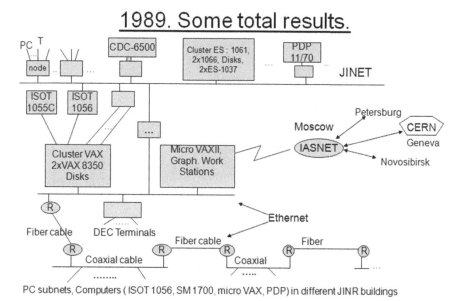

Fig. 5. Use of Ethernet connections at the end of the 1980s

It was time to dismount the BESM-6 and connect new servers to the Ethernet. A cluster of two VAX-8350 that contained shared disk memory (about 12 Gbyte), cartridge units (200 Mbyte), and high density (up to 6250 bit/inch) magnetic tape units. Additionally, the system contained the powerful MEGATEK/WIZZARD 7555 station with three-dimensional graphics.

6 Activities during the 1990s

At the very beginning of the 1990s, the throughput of the Russian terrestrial external channels was inadequate for territorially distributed information exchange and processing in high-energy physics, which had practical limits of 32K bits/sec. This is why two direct satellite channels were organized at JINR and started to be exploited in 1994 for access to the DFN (Germany) networks and the High Energy Physics NETwork (HEPNET) with an entry point in Italy and a terrestrial channel to CERN. The throughput of each channel was up to 64K bits/sec; we used our own antenna and the Soviet "RADUGA" satellite for first channel and the terrestrial link to a nearby Cosmic Communication Station "DUBNA" and international satellite INTELSAT for second channel. We had a plan to give some other Soviet scientific institutes such as the IHEP in Protvino the possibility to share this second channel with JINR. It description appears in [2] and shown in Figure 6. Of course, terrestrial channels, with the same throughput as satellite channels, have an advantage in many cases and when new facilities to increase their throughput appeared (up to 1 Gbit/sec nowadays), we had taken the opportunity and ended the use of satellite channels.

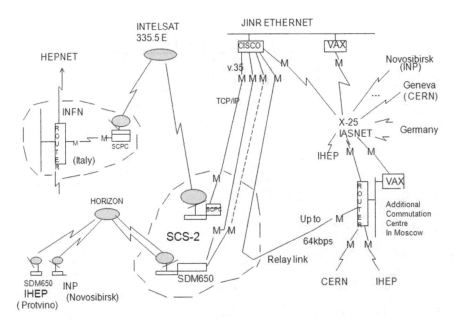

Fig. 6. Shared resources during the 1990s

6 Conclusion

In conclusion, we have seen some close perspectives in the use of distributed systems by a community of scientific organizations. They are, in particular, the final realization of an international project called the Worldwide LHC Computing Grid (WLCG) [6]. Eight institutions in the Russian Federation (in Moscow, Moscow district, St. Petersburg) became the participants in this project. The JINR Scientific Research Institute for Nuclear Physics of Moscow State University (SRINPofMSU), ITEP, IHEP, IAM, Research Scientific Centre of Kurchatov`s Institute (RSC KI), Petersburg`s Institute of Nuclear Physics (PINP) and the Institute of Mathematical Problems in Biology (IMPB) as institutes of Russian Academy of Science.

Acknowledgments. I am grateful to several people (especially to Georg Trogeman, Alexander N. Nitussov and Wolfgang Ernst) who organized the publication of the book mentioned in [3,4] that contains much information concerning computing and networking in Soviet Union. Thanks also go to the Ministry for Schools, Education, Science and Research of the German state North-Rhine Westfalia and to the Academy of Media Arts Cologne for supporting this publication.

References

1. Foster, I.: What is the Grid? A Three Point Checklist,
   http://www.gridtoday.com/02/0722/100136.html
2. Shirikov, V.P.: Matematicheskoye obespechenie vychislitelnyh kompleksov i setei. (The Mathematical Support of Computation Complexes and Networks). In: Programmirovaniye (3). Nauka, Moscow (1991)
3. Apokin, I.A.: The Development of Electronic Computers in the USSR. In: Computing in Russia. The History of Computer Devices and Information Technology revealed. Vieweg, Germany (2001)
4. Shirikov, V.P.: Scientific Computer Networks in the Soviet Union. In: Computing in Russia. The History of Computer Devices and Information Technology revealed. Vieweg, Germany (2001)
5. The CERN openlab: a novel testbed for the Grid. In: Cern Courier (October 2003)
6. Memorandum of Understanding for Collaboration in the Development and Exploitation of the Worldwide LHC Computing Grid, http://lcg.web.cern.ch/LCG/C-RRB/MoU/WLCGMoU.pdf

Automatic Digital Computer M-1 of the I.S. Brook Laboratory

T.M. Alexandridi[1] and U.V. Rogachov[2]

[1] The Moscow State Technical University (MADI)
alexandridi@mail.ru
[2] Scienntific Research Institute of Computer Systems (SRICS)
rogachev25@rambler.ru

Abstract. This article describes the history of creation of the first digital computer in Russia. It started in the electronic systems laboratory of the Energy Institute of the USSR Academy of Science, led by I.S. Brook. The first developer was N.Y. Matyukhin. The computer M-1 consisted of arithmetic unit, main program sensor (control device), internal memory of two types (fast one – on cathode-ray tubes and the slow one – on magnetic drum), and an input/output device using telegraph alphanumeric equipment. By the end of 1951, for the first time in the world, the M-1 started working in experimental mode with fast-acting storage device made on common LO-737 cathode-ray tubes. The computer successfully passed complex tests and was switched to operational mode. For the first 1.5 years, it had been the only computer in the USSR.

Keywords: I.S. Brook, N.Y. Matyukhin, first computer in Russia, M-1.

1 Introduction

In the December of 1951, the electronic systems laboratory of the Energy Institute of the USSR Academy of Science, led by I.S. Brook, issued a scientific and technical report called the "Automatic digital computer M-1". It was the first scientific document about the creation of native electronic computer in the USSR. The machine had successfully passed the required tests and soon, they had switched it to operational mode.

The beginning of I.S. Brook's scientific researches on issue of digital computers dates on 1948. He was the first person in the USSR (together with B.I. Rameev) to develop the design of a digital computer with hard program control. First certificate on invention of "Electronic computer with common bus" appeared in December 1948.

A resolution of the USSR Academy of Science presidium on the start of the development of the M-1 was published 22 April 1950. I.S. Brook was in charge to form a command of developers. The first developer was N.Y. Matyukhin, a young specialist who just graduated with honors in radio engineering from the Faculty of the Moscow Energy Institute. Brook familiarized him with the main concepts of construction of the electronic computer, and soon they together developed a future computer structure, its main characteristics, and specific solutions of many technical

J. Impagliazzo and E. Proydakov (Eds.): SoRuCom 2006, IFIP AICT 357, pp. 46–49, 2011.

situations. Later on, Matyukhin with permanent support of Brook practically carried out all the functions of the head designer. Below is a fragment from memoirs of Nikolay Yakovlevich:

"I'll make an effort to revive pictures of our work under I.S. Brook's direction, reproduce the atmosphere of the first working years in the sphere of computer engineering."

2   The M-1 and Its Development Team

The formation of group and the commencement of work on the AEC (automatic electronic computer) M-1 began in the summer of 1950. Brook enrolled a command of young specialists from the radio engineering faculty consisted of seven people:

- o   Two junior research officers – A.B. Zalkind and N.Y. Matyukhin
- o   Two students engaged on their degree thesis – T.M. Alexandridi and M.A. Kartsev
- o   Three technicians – Y.V. Rogachev, R.P. Shidlovsky, and L.M. Zhurkin

A first assignment from Isaac Semenovich was to develop a valve for a triple-entry summarizer, as I recall. The second assignment was to produce a standard working table. The third assignment from the group manager was the development of the AEC M-1.

Serious difficulties occurred during the development of the M-1. The project realization was in jeopardy due to a complete lack of component parts. Isaac Semenovich suggested a peculiar way to obtain the necessary material by using the storehouses from the war years. The resulting elements of the project consisted of the following ideas and "trophies":

- o   combinations of small items made of components from very different origins
- o   only two types of electronic tubes
- o   vitriolic rectifiers of electronic measurement equipment
- o   cathode-ray tubes for the oscillograph
- o   Trophy teletype from Vermacht's general material

Some of Brook's guiding principles included:

- o   deep understanding of goals, simplicity and vividness of argumentation;
- o   no "excoriations" on failures;
- o   respectful treatment to performers.

The AEC M-1 consisted of arithmetic unit, main program sensor (control device), internal memory of two types (fast one – on cathode-ray tubes and the slow one – on magnetic drum), and an input/output device using telegraph alphanumeric equipment. In the autumn of 1950, the N.Y. Matyukhin "command" reached completion.

Matyukhin and Rogachev were in charge of developing the arithmetic unit and the logical element system. Kartsev and Shidlovsky oversaw the development of the main program sensor system. Matyukhin and Zhurkin were responsible for the storage device on magnetic drum and T.M. Alexandridi was in charge of the storage device on the cathode-ray tube. Zalkind and D.U. Ermochenkov developed the input/output device system. V.V. Belynsky oversaw the power development and A.I. Kokalevsky supervised the construction design. Matyukhin was in charge of the complex tuning of the computer, its technology, and its regiment of performance testing.

3 Designing the M-1

Fundamentally, new technical solutions were proposed and realized during design and development of M-1, especially the double address command system, which we used widely later on in native and foreign computer engineering. Later Matyukhin wrote about this solution as follows.

> "Command system choice was very tricky case – at that time triple address system was generally accepted and thought to be mostly natural. This system required rather high digit rate of register equipment and memory. Our limited possibilities stimulated researches of more efficient solutions. As it sometimes happens in dead end situations, occasion helped. At that time I.S. Brook invited young mathematician Y.A. Shreider. Shreider assimilated the basis of programming, directed our attention on the fact that in many formulas of approximate calculation operation result is the operand for next operation step. That was the start of the first double address system. Our proposals were approved by I.S. Brook, and AEC M-1 became the first double address system".

For the first time in world practice of computer design, we built diode logical elements on semi-conducting elements. Matyukhin estimated the significance of this step in his memoirs:

> "One of the principal solutions, that, on my mind, predetermined the success of our first computer and short creation time, was the line, accepted by Brook on usage of semi-conducting elements. At that time, they appeared in our industry only in compact vitriolic rectifiers, issued for measurement equipment purposes. Brook arranged the issue of special modification of those rectifiers in proportions of common resistor, and we created the set of typical schemes. The creation and assembling of blocks started in laboratory workshop, and less than in one year computer began to "breathe".

Serious problems had arisen during design of electrostatic memory. We all knew that the USA and the USSR had developed special cathode-ray tubes – potential scopes. However, they were inaccessible for us. I.S. Brook determined to use common cathode-ray tubes LO-737, used in oscillographs.

Around that time in September of 1950, a new student engaged doing a degree thesis from the radio engineering faculty appeared in laboratory; her name was Tamara Alexandridi. Brook suggested carrying out the degree thesis in "electrostatic storage devices", intended for AEC. She conducted the first experiments on cathode-ray tubes of the common oscillograph and other additional measurement equipment.

These experiments showed that indeed, they could achieve the information memorization effect on cathode-ray tubes. Thus, the design of electrostatic storage device for M-1 started. In the spring of 1951, Alexandridi wrote and defended her thesis called, "Electrostatic storage devices". At the same time, the invention of the electronic memory came into existence. By the end of 1951, for the first time in the world, the M-1 started working in experimental mode with fast-acting storage device made on common LO-737 cathode-ray tubes.

4 M-1 Characteristics

Work on M-1 tuning ended in the autumn of 1951. By December of that year, the computer successfully passed complex tests and was switched to operational mode.

The M-1 had interesting characteristics. We delineate them here.

- o   Scale of notation – binary.
- o   Number of binary classes – 25.
- o   Command system – double address.
- o   Internal memory size:
       On electrostatic tubes – 256 addresses,
              On magnetic drum – 256 addresses.
- o   Operational speed:
       With slow-acting memory – 20 operations/sec,
       With fast-acting memory:
              Composition – 50 mksec,
              Multiplication – 2000 mksec.
- o   Electronic valves quantity – 730.

5 Conclusion

Academicians A.N. Nesmeyanov, M.A. Lavrientiev, S.L. Sobolev, and A.I. Berg became acquainted with the work of AEC M-1. One of the first people to solve atomic objectives on the computer was Sobolev, who was an assistant on scientific work in Kurchatov's institute.

For three years, the machine had been running in operational mode and for the first 1.5 years, it had been the only computer in the USSR. Only one copy of the M-1 existed. However, its architecture and many of its principal schematic solutions became a foundation for the development of a series of new machines such as the M-3, the "Minsk", and the "RAZDAN".

The M-1 creators became great specialists in computer engineering and noticeably contributed to computer development; they spearheaded many different scientific, educational, and industrial collectives. Their works had been highly recognized by academic degrees, honorary titles, and governmental awards. In the end, N.Y. Matyukhin became a Corresponding Member of USSR Academy of Science. Doctor of Science, Professor M.A. Kartsev became the director of Science and Research Institute of Computing Complexes. They both were the principal designers of large computers in the USSR.

Conception of New Generation Computer Systems – The Last Large-Scale Initiative in Computers of the COMECON Countries: A Glance after Twenty Years

Victor Zakharov and Yuri Lavrenjuk

Institute of Informatics Problems of the Russian
Academy of Sciences, Moscow, Russia
vzakharov@ipiran.ru, lavr@voxnet.ru

Abstract. The conception of a new generation of computer systems was developed by a team of scientists from USSR Academy of Sciences, academies of USSR republics, academies of COMECON countries in 1984-85. The conception contained main directions of computing technology and informatics development in the socialist countries. Additionally, it prescribed the main directions of research and development that should be performed to obtain new qualities of information and computer systems. The conception was supposed to be implemented within the framework of ten integrated scientific projects performed by the international research teams. The structure of these projects and the main results of their implementation are presented. A glance on these projects from today positions is also presented.

Keywords: New generation computer systems, scientific project, research team, COMECON countries.

1   Introduction

In December 1983 in Sofia, the capital of Bulgaria, according to the recommendation of the meeting of the Academies of Sciences of the Socialist Countries representatives, the First Meeting of the Coordination Council for Computing Technology and Informatics of the Academies of Sciences of the Socialist Countries (CCCTI) took place. Academician Evgeny Velikhov, vice president of the USSR Academy of Sciences, headed this council. The representatives decided at this meeting to organize the Provisional Work Team (PWT) for working-out the "Conception of a new generation of computer systems" in the socialist countries. The proposals provided for directions of fundamental and applied research in computing technology and informatics.

Representatives of eleven Academies of Sciences of the socialist countries participated in PWT. They included the People's Republic of Bulgaria, Hungarian

J. Impagliazzo and E. Proydakov (Eds.): SoRuCom 2006, IFIP AICT 357, pp. 50–63, 2011.

People's Republic, Socialist Republic of Vietnam, German Democratic Republic, Korean People's Democratic Republic, Republic of Cuba, Mongolian People's Republic, Polish People's Republic, Socialist Republic of Romania, Union of Soviet Socialist Republics, and the Czechoslovak Socialist Republic. Note that only three of these countries exist now with the same names (Vietnam, Korea, and Cuba). The director of the Institute of Informatics Problems of the USSR Academy of Sciences, Boris Naumov (at that time a corresponding member of the USSR academy, elected in 1984 a full member of the USSR academy), was appointed as the Chairman of PWT.

During 1984-85, a team of scientists from the USSR Academy, the academies of the Soviet republics, the academies of COMECON countries, after intense common activity, developed the "Conception of a new generation of computer systems" (CNGCS), which was approved in June 1985 in Prague on the third meeting of CCCTI [1]. That conception represented in some sense the answer of the socialist countries on the ESPRIT program of EEC and on the Japanese program of the fifth-generation computers. The conception contained main directions of computing technology and informatics development in the socialist countries for the period up to 2010. Additionally, it prescribed the main directions of research and development that should be performed to obtain new qualities of information and computer systems. The conception became the important component of the Integrated Program of Scientific and Technical Progress of the COMECON countries, the base for carrying out unified technical policy in academies of sciences and industry of the socialist countries.

The conception was supposed to be implemented within the framework of ten integrated scientific projects (ISP) performed by the international research teams. They included:

"Systems of knowledge processing" (ISP-1)
"Image processing and computer graphics systems" (ISP-2)
"CAD for computer systems" (ISP-3)
"Computer Networks" (ISP-4)
"Systems of personal computers" (ISP-5)
"Fault-tolerant systems" (ISP-6)
"New external storage devices" (ISP-7)
"Software engineering for new generation computer systems" (ISP-8)
"New algorithms and architectures for information processing" (ISP-9)
"Informatics for education" (ISP-10)

To guide the realization of the conception, CCCTI had arranged as its work institution the Commission on the New Generation Computer System (CNGCS) and academician Boris Naumov was appointed its Chairman. The main task formulated for CNGCS was to provide a practical realization of all projects using all efficient forms of international collaboration such as establishing international institutes and laboratories, arranging international conferences and symposia, creating a special

journal, and publishing proceedings. Due to these decisions, in 1988 they launched
the publication "Journal of new generation computer systems" by the "Akademie-
Verlag Berlin" publishing house at the Central Institute for Cybernetics and
Information Processes of the Academy of Sciences of the G.D.R. The course of the
Conception fulfillment appeared in the issues of this journal. This journal was edited
up to 1991 when the activity on the conception realization was practically terminated
due to the well-known events in most of the participating countries; fourteen issues of
the journal were published.

We will now try to present in a short form the goals and content of the integrated
scientific projects and give some comments about the confirmation and correctness of
the ideas included to the conception, its achievements, and its hopes.

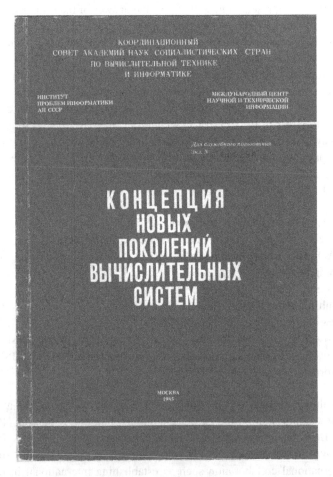

Fig. 1. Conception (photo of cover)

Fig. 2. Journal (photo of two cover pages of the journal)

2 Systems of Knowledge Processing (ISP-1)

Responsible Country: Czechoslovakia
Participating Countries: Bulgaria, Hungary, Germany, USSR
Supervisor: Academician Ivan Plander (Czechoslovakia)

2.1 Working Goal

A step-by-step creation of artificial intelligence systems providing efficient implementation of the knowledge processing functions and basing on architecture with high level of concurrency and distributed data bases. Development by 1995 a knowledge processing computer with special concurrency means. It was planned to perform main joint researches in the base laboratory on artificial intelligence in Bratislava (Czechoslovakia). The ISP-1 included six projects described below.

2.1.1 Project "Expert Systems"
Lead by Hungary with USSR and Czechoslovakia as participants.

Goal: Creation of a modular "empty" hardware system with a flexible architecture (expert meta-system)

Project Result: It created the universal environment for constructing expert systems for diagnostics and planning tasks.

2.1.2 Project "Main Tools for Knowledge Processing"
Lead by Germany with Hungary, USSR, and Czechoslovakia as participants.

Goal: A systematic creation of hardware and software means and specialized architecture models for knowledge processing.

Project Result: Development of the conception of the first release of a knowledge-processing computer.

2.1.3 Project "Systems of Knowledge Processing Computers"
Lead by Czechoslovakia with Germany, Hungary, and USSR as participants.

Goal: Creation of efficient implementation for knowledge processing functions on system architecture with high level of concurrency and with the use of relational data base facilities.

Project Result: Developed relational data base computer and knowledge base system of SIMD type; a sample of knowledge processing system (SOZ-1) was created.

2.1.4 Project "Distributed Knowledge Base Systems"
Lead by USSR with Hungary, Germany, and Czechoslovakia as participants.

Goal: Development of methods and tools for cooperative problem solving, using systems of distributed knowledge bases and mechanisms of distributed solutions of tasks.

Project Result: Created a "synthesis" knowledge integration system. Actively growing DATA GRID system represents in some degree an example of the tasks formulated in the project realization.

2.1.5 Project "Man-Machine Communication on Natural Language"
Lead by USSR with Germany and Czechoslovakia as participants.

Goal: Development of algorithms and methods for hardware-software implementation of input/output of natural language written texts and speech (partially, safe understanding of continuous speech without training on definite speaker).

Project Result: Developed a number of decent quality systems for output (pronouncing) of written texts. Understanding of written natural texts and especially continuous speech understanding happened to be too difficult tasks and are still not implemented in the full volume.

2.1.6 Project "Data Base Computer"
Lead by USSR with Germany and Czechoslovakia as participants.

Goal: Development of specialized computers for data base systems control.

Project Result: Limited progress on this effort.

2.2 Result of Performed Work

The projects developed a model of a database computer. Currently, a number of different companies in some countries are manufacturing database computers, which are operating within big information systems.

3 Image Processing and Computer Graphics Systems (ISP-2)

Responsible Country: Germany
Participating Countries: Bulgaria, Hungary, Poland, Romania, USSR, Czechoslovakia
Supervisor: Professor V. Vilhelmi (Germany)

3.1 Working Goal

The goal of this endeavor was the development of unified hardware and software facilities and standardized design solutions for the creation systems for image processing and computer graphics. The expectation by 1990 was the creation of a highly productive unified components and system of image processing with modular architecture. The ISP-2 supposed activity within three projects described below.

3.1.1 Project "Development of Digital Image Processing Systems"
Goal: Creation of universal systems to be used as the experimental basis for algorithms and solutions development; creation of unified components for specialized peripheral devices; prototypes development of highly productive components basing on new algorithms and new architectural solutions.

3.1.2 Project "Development of Standard Applications for Automated Image Processing and Analysis"
Goal: Creation of real time image processing systems (for industrial automation and scientific researches); creation of systems for 3-dimensional scenes analysis using knowledge bases; creation of "turnkey" systems for selected applications.

3.1.3 Project "Creation of Computer Graphics Systems"
Goal: Development of efficient algorithms and basic components for three-dimensional graphics processing; development of standard solutions for CAD systems in mechanical engineering, aviation engineering and other industrial areas.

3.2 Result of Performed Work

Within the integrated project, the results included a search of efficient specialized hardware solution oriented on semi-custom and custom VLSI that came to nothing more than experimental samples and models and they did not receive support from industry. Methods of software implementation were applied in some application systems based on general-purpose computers. Partially, it created a standard system for image processing on PC, a sample of intellectual controller for video databases, and a sample of system for images obtained from space processing.

Later, as a continuation of these works, the system for images analysis, understanding, and recognition named "Black square" came into existence in the Cybernetics Scientific Council of the Russian Academy of Sciences. Currently, powerful systems of 3-D graphics based on specialized graphical controllers are actively used in many applications.

4 CAD for Computer Systems (ISP-3)

Responsible Country: USSR
Participating Countries: Germany, Bulgaria, Poland, Czechoslovakia
Supervisor: Doctor of science, Efim Oihman (USSR)

4.1 Working Goal

The activity sought to create an integrated distributed CAD for computer systems, covering all stages of design from the level of system structure up to simulation, verification and testing of integrated circuits, image generators and testing procedures control, development of designer's documentation, control of manufacturing equipment and production planning. The system should allow describe and simulate IC's with the integration level up to 10^8 transistors on a chip using a high-level language of a "Silicon compiler" type that included a knowledge base and expert system. It was also to develop an experimental CAD system by 1990. ISP-3 included three projects described below.

4.1.1 Project "Systems for Description and Modeling Architectural Level of Computers"
Lead by Poland with USSR and Czechoslovakia as participants.

4.1.2 Project "VLSI and ULSI CAD Systems"
Lead by USSR with, Germany and Czechoslovakia as participants.

4.1.3 Project "CAD System for Computers"
Lead by USSR.

4.2 Result of Performed Work

A first version of a silicon compiler operating on 32-bit computer was developed and tested. Currently, there is a number of CAD computer systems, which include all functions described in the project. It provides all time parameters planned.

5   Computer Networks (ISP-4)

Responsible Country: Germany
Participating Countries: Hungary, Bulgaria, USSR, Czechoslovakia
Supervisor: Doctor of science, V. Heymer (Germany)

5.1 Working Goal

Development of methodology, standards on protocols, unified hardware and software components for the creation of local and global computer and data transmission networks, development tools for computer networks design, and tuning and exploiting automation. ISP-4 included four projects described below.

5.1.1 Project "Local Networks"
Lead by Bulgaria with Hungary, USSR, Czechoslovakia, and Germany as participants.

Supervisor: Professor Kiril Boyanov (Bulgaria).

Goal: Develop systems of protocols and create local networks providing transmission of data, sound, and images between different types of computers and devices, integrated into the united system; develop architecture of gateways between different networks taking into account upcoming international standards.

5.1.2 Project "Hardware and Software Components for Network Processors"
Lead by Germany with USSR and Czechoslovakia as participants.

Supervisor: Doctor of science, V. Heymer (Germany).

Goal: Create a profitable, efficient, and safe multiprocessor computer system to perform functions of package switching in global data transmission networks (parameters: up to 1000 communication lines, carrying capacity: up to 1000 packages/sec, full operating time: 5-10 years); perform research of hardware and software solution for connecting computers and intelligent terminals to global data transmission networks with the goal to create a family of communication devices.

5.1.3 Project "Network of Intelligent Terminals"
Lead by Czechoslovakia with USSR and Germany as participants.

Supervisor: Doctor of science, Miroslav Novak (Czechoslovakia).

Goal: Create an integrated network of terminals, independent on types of included hardware devices, which permit simultaneous connecting a great number of intelligent terminals (different types of personal computers, mini-computers, special processing devices for telecommunication equipment, for experiment control and so on) to mainframes and computer networks via communication processors.

5.1.4 Project "Distributed Data Processing"

Lead by Hungary with Bulgaria and Germany as participants.

Supervisor: P. Bakoni (Hungary).

Goal: Develop principles of task distribution between elements and create distributed data processing systems in computer networks on the principles of ISO open systems; implement as the experiment a teleconference system.

5.2 Result of Performed Work

Some basic researches on conceptual level were performed, a number of LANs and package switching devices, network of intellectual terminals with communication processor were designed and provided for manufacturing.

6 Systems of Personal Computers (ISP-5)

Responsible Country: USSR
Participating Countries: Bulgaria, Hungary, Germany, Korea, Mongolia, Poland, and
    Czechoslovakia
Supervisor: Doctor Igor Landau (USSR)

6.1 Working Goal

Develop architectural and software solutions for prospective personal computers (PC) and their peripheral devices oriented on artificial intelligence tasks processing – both in the role of intellectual terminals of computer systems and as stand-alone intellectual workstations.

The project was based on the forecast about occurring everywhere usage of PCs and possibility of their connection to information and computer networks providing access to general use data and knowledge bases. Methods providing software portability have been investigated. They developed: tools for architecture description; structure and components of software for new generation PC; new man-machine interfaces providing input of tasks and requests to a system on language close to natural one.

6.2 Result of Performed Work

Models of 32-bit PC with microprocessor compatible with I80386 have been developed and delivered to the plants for manufacturing; microprocessors with RISC architectures and transputers were also developed. They also developed such software as a universal operating system UOS (those time it exceeded all existed on market OS for PC in some characteristics), a portable OS of UNIX type, standard packages of application programs for PC.

Passed time confirmed the correctness of the estimation regarding the role and place of the PC in new generation systems and surpassed the most audacious forecasts about the scale of PC expansion and applications.

7 Fault-Tolerant Systems (ISP-6)

Responsible Countries: USSR, Germany
Participating Countries: Bulgaria, Czechoslovakia
Supervisors: Academician Boris Naumov (USSR) and professor Dieter Hammer
(Germany)

7.1 Working Goal

Develop experimental samples of new generation fault tolerant computer systems, with ability to reorganize automatically, practically without breaks in operating their structure with quick replacing of fault components after faults detecting. Develop methods to achieve "absolute" (during all system life period) reliability alternative to traditional methods of dubbing, tripling, etc. Expected ways of solving these problems include usage of self-timed schematics on the LSI and VLSI, fault-tolerant interfaces and system software supporting functions of failure parrying on the structural levels of complexes and systems.

7.2 Result of Performed Work

A number of works on the level of resolving conceptual items had occurred. Actuality of the formulated problem did not disappear up to present time. The investigations in self-timing technology only recently achieved practical stage – samples of simple microprocessors in chip designed within self-timed schematics were manufactured and tested. They demonstrated the correctness of theoretical proposals regarding considerable expansion of the operating environment area and efficiency of fault-tolerant systems creation basing on self-timed schematics.

8 New External Storage Devices (ISP-7)

Responsible Country: Bulgaria
Participating Countries: Hungary, Germany, USSR, Czechoslovakia
Supervisor: Professor J. Kasabov (Bulgaria)

8.1 Working Goal

Development of new methods for recording and storing information on external storage devices (ESD), providing increase of capacity and reliability. Development of the methods for information protection in ESD and implementation of these methods in VLSI set. ISP-7 included five projects outlined below.

Project "External storage devices based on multi-layer medium with optical writing and electrical amplification; integral-optical modular ESD; hierarchical holographic ESD; associative ESD".
Project "ESD based on multi-layer silicon structures".
Project "ESD based on bubble memory".
Project "Devices of ultra- thick magnetic recording of digital information".
Project "Information protection in ESD".

8.2 Result of Performed Work

Some investigations and developments have been performed on the base of the Laboratory of new ESD in Bulgarian Academy of Sciences. The works performed within this ISP did not play sizeable role as they were behind the leading world investigations in this area.

Currently, the achieved parameters of ESD capacity and reliability greatly exceeded all forecasts and expectations announced more than twenty years ago. Thus, for example, it was formulated a task to achieve capacity 1 Gbytes on one magnetic disc; now 100 Gbytes capacity is usual for HDD in PCs. It was a task to achieve 1–5 Mbyte on semiconductor external memory, now flash-memory devices with more than four Gbytes are available.

9 Software Engineering for New Generation Computer Systems (ISP-8)

Responsible Country: USSR
Participating countries: Bulgaria, Germany, Mongolia, Czechoslovakia
Supervisor: Doctor Valentin Semik (USSR)

9.1 Working Goal

Creation of scientifically substantiated programming technology for new generation professional personal computers, covering the whole "life cycle" of software product and setting up the basis for industrial software and software documentation manufacturing; development of new methods for increasing reliability of application computer programs. ISP-8 included three projects outlined below.

   Project "Experimental system of automated design, production, and support of software for new generations professional personal computers".
   Project "Development of the methods for software reliability improvement".
   Project "Technology of computer system architecture and software design".

9.2 Result of Performed Work

There were created instrumental-technological complexes of object-oriented programming for graphical systems development, for concurrent and pipeline calculations design and implementation.

It is reasonable to note that within this project it was paid special attention to the difference between *technology of programming* (tools, supporting program design, and coding in computer systems) and *technology of tasks solution* (designated for the end users of application systems). Such approach is obvious now and it is widely used in computer software development.

10   New Algorithms and Architectures for Information Processing (ISP-9)

Responsible Countries: USSR, Bulgaria
Participating Countries: Germany, Czechoslovakia
Supervisors: Academicians Vladimir Pugachev and Alexander Samarsky (USSR) and
  academicians S. Markov, professors Kiril Boyanov and J. Mikloshko (Bulgaria)

10.1   Working Goal

The activity had four specific goals described as follows.

o   Investigation and development of the common conceptual basis for evolution of mathematical modeling and computer experiment methods for solving tasks destined to creation of perspective computer hardware, its element base and software;
o   Development and investigations of methods for automated compilation on computers equations for statistical characteristics in the tasks of stochastic systems processes analysis and synthesis, implementation in practice corresponding application packages;
o   Development of algorithms and architectures for new types of computer systems including algorithms for concurrent calculations, implementation in practice corresponding application packages;
o   Creation of multi-processor systems of traditional and non-traditional architecture basing on concurrent data processing.

10.2   Result of Performed Work

Based upon a great number of performed fundamental researches there were developed application packages (including such as for the big tasks of linear algebra, stochastic systems, processes in electron-hole plasma, heat load on computer components, etc) oriented toward users and non-professional programmers; they developed special language tools destined to create software for data driven computers.

Currently, the process of development and accumulation of different methods and algorithms for different type programs solving is actively continuing. Most of the ideas formulated in these projects found their realization in large modern mathematical software systems.

11   Informatics for Education (ISP-10)

Responsible Country: Bulgaria
Participating Countries: Hungary, Germany, Mongolia, USSR
Supervisor: Academician Blagovest Sendov (Bulgaria)

11.1 Working Goal

Creation of algorithmic, software and hardware means for informatics in education based on new generation computer systems (at first – on professional and educational personal computers) and carrying out large-scale experiments on means and methods of information technologies application in education. The main direction for this activity included the following.

o Investigation of necessary modifications in learning programs and methodologies to bring them in correspondence with potentials of new information technologies;
o Development of learning programs and manuals on informatics for all levels of education;
o Development of software means for interaction with computer (language for primary education, information-logical languages, structural programming languages, languages for robots controlling and so on);
o Creation of the models for teaching processes and tasks solving;
o Integration of different means and methods for construction learning process models using methods of cognitive psychology and artificial intelligence;
o Investigation and development of methods and tools for certification, replication and support of software for the area of educational informatics;
o Development of software tools for learning courses and pedagogical programs preparation;
o Creation of dialog teaching and consulting systems using methods of logical deduction, knowledge bases and modeling basing on the conception of computational experiment;
o Development of architecture of simulator complexes for teaching popular professions bind with computer (microprocessor) devices manufacturing and application;
o Organization of experimental centers of educational informatics for testing and tuning scientific and organizing forms of the research results familiarization.

11.2 Result of Performed Work

A considerable number of methodologies and software means ensured new information technologies usage in different levels teaching processes were developed. The active implementation of information technologies into the learning process took place in many countries. Special research Institute for Information Technologies in Education was established by the UNESCO (in Moscow, Russia).

Investigations in the area of wide implementation of information technologies to education process are continuing also at present in the most part of formulated in the project directions. In general the correctness of the formulated ides have been confirmed during passes twenty years.

12 Conclusion

From the above, we can see that in the process of its activity, the CNGCS headed by academician Boris Naumov succeed in detecting the main stream among many details and placed the emphasis on creation of *new generation computer systems*, not only on *computer devices*. Additionally, individual projects have been interconnected by joining them up in integrated scientific projects, which also have been closely interconnected. Finally, almost everything formulated in the conception projects looks today as obvious – exactly in such way modern information and computational systems are created. However, it is necessary to take into account that these projects have been prepared more than twenty years ago, where at that time many things were not so obvious as they are today; PCs, the internet, and mobile communications were not widespread technologies at that time.

Reference

[1] Conception of New Generation Computers. International Center of Scientific and Technical Information, Moscow (1985) (in Russian)

SM EVM Control Computer Development

N.L. Prokhorov[1] and G.A. Egorov[2]

[1] Supervisor of Studies, INEUM, Moscow, Russia
prokhorov_n@ineum.ru
[2] Deputy General Director, INEUM, Moscow, Russia
egorov_g@ineum.ru

Abstract. The main stages in the development of the Institute of Electronic Control Computers during fifty years of its existence are considered in this paper. The basic models of controlling computer systems, developed at the institute are presented. Creating and large-scale industrial production of small-computer system SM EVM have founded the technological base of automation control and information processing in various branches of industry of the country.

Keywords: INEUM, control computers, SM EVM, real-time control.

1 Introduction

One of the major development paths of the national computer engineering is associated with the Institute of Electronic Control Computers (INEUM), organized by its first director, one of the pioneers of the national computer engineering, Corresponding Member of the USSR Academy of Sciences I.S. Bruk (1902 - 1974).

As the school of Academician S.A. Lebedev was oriented toward designing maximum-throughput computers for each generation of elements, I.S. Bruk's conception from the very beginning was vectored to the class of small and medium computers for which the cost/performance index and balanced characteristics are extremely important. It is interesting to state that both S.A. Lebedev and I.S. Bruk proceeded to design digital computers on the background of the experience obtained by using the analog computers in the area of power engineering and deep understanding of its disadvantages.

2 Virtues of I.S. Bruk

The first Soviet inventor certificate for the digital computer was received by I.S. Bruk and B.I. Rameev with priority of December 1948. The automatic digital computer M-1 was designed at the Laboratory of Electrical Systems of the Energy Institute of the USSR Academy of Sciences in 1950–1951 under the guidance of I.S. Bruk. It was brought into service in 1952, several months later than the MESM computer designed by S.A. Lebedev in Kiev. The main concepts of M-1 were formulated by I.S. Bruk and N.Ya. Matyukhin, who was then a young graduate of the Moscow Power Institute

J. Impagliazzo and E. Proydakov (Eds.): SoRuCom 2006, IFIP AICT 357, pp. 64–73, 2011.

(MEI) and future Corresponding Member of the USSR Academy of Sciences. The M-1 realized the two-address instruction system and some important decisions concerning selection of logic and circuitry of the digital computers, which played an important role in the future development of the national computer engineering.

It is important to notice here that the designers of MESM and M-1 arrived to the classical design of their digital computers based on the stored-program (von Neumann) architecture independently of each other. The works of the American scientists were known in the USA since 1946, but first published in an abbreviated form in 1962.

In the laboratory of I.S. Bruk in 1952, a group of MEI graduates headed by M.A. Kartsev developed the M-2 computer. Working at approximately the, same speed as the "Strela" computer, M-2 had four times less number of electronic tubes, consumed seven to eight times less power and occupied one-tenth the area. These achievements were due to the use of conventional oscillographic cathode-ray tubes as main-memory elements and semiconductor diodes in the logic circuits. It seems that it was for the first time that at upgrading M-2 in 1953-1956 M.A. Kartsev realized the idea of shortened instruction addresses (with switching of the memory areas) and shortened operation codes as means of coordinating the instruction and number formats. This idea preceded the methods of generating the execution addresses in the computers of the second and third generations.

Based on the experience gained in the work on M-1 and M-2, I.S. Bruk formulated the concepts of "small-sized" computers. The first solution of this problem was embodied in M-3 designed 1956 in collaboration with LUMS of the USSR Academy of Sciences (I.S. Bruk) and NIIEP (Academician A.G. Iosif'yan). The M-3 philosophy was formulated by I.S. Bruk, N.Ya. Matyukhin, V.V. Belynskii, B.M. Kagan, and V.M. Dolkart. M-3 served as a prototype of two commercial families of domestic computers realized by the designers of M-3: "Minsk" (G.P. Lopato, later Corresponding Member of the USSR Academy of Sciences) and "Razdan" (B.B. Melik-Shakhnazarov, Institute of Mathematics of the Armenian Academy of Sciences from which the Yerevan Institute of Mathematical Machines was later separated).

3 The INEUM Era

Establishment of INEUM was preceded in 1957 by formulation by I.S. Bruk of the scientific problem "Development of the Theory, Principles of Design and Application of the Special-Purpose Computing and Controlling Machines." The problem report compiled by an expert team headed by I.S. Bruk was published by the USSR Academy of Sciences in 1958 in the serial edition "Problems of Soviet Science". The report indicated to the significance of the control machines for the national economy and, for the first time, presented and substantiated the main directions of fundamental and applied studies in the field of production automation and plant control by special-purpose and control machines.

In the course of time, the notion of special-purpose control and computing machines as formulated in the report with regard to the first-generation computers changed its initial significance owing to the rapid progress of electronics. At the same time, the I.S. Bruk's notion of the "control computers" which are distinguished from the general-purpose computers in the nature of their relation with the control plant, higher reliability, real-time operation, ability to work in unfavorable industrial environment, and so on retained its value and was further developed.

In 1957 a team of I.S. Bruk's disciples headed by N.Ya. Matyukhin joined a defense research institute to design digital computing facilities for the air defense systems. In 1957, while staying with INEUM, N.Ya. Matyukhin independently of M. Wilkes proposed the principle of the microprogram computer control that required detailed working out of the algorithms of arithmetic operations, their decomposition into elementary micro-operations and prototyping of the circuits of the microprogram computer with microprograms stored in the read-only memory. This principle was then widely used in computers both in INEUM and elsewhere by the designers of control and other computers.

In 1958-1964 INEUM developed the control computer M4 (M4-M, M4-2M) intended for real-time control of a radar complex (Radiotechnical Institute of the USSR Academy of Sciences, Academician A.L. Mints) incorporated in the radioelectronic system of Earth satellite observation. M4 was one of the first domestic computers based on the second-generation circuitry. Although the decision about commercial production of M4 (M4M) was made in 1962 after successful tests on the working prototype of the radar complex, M.A. Kartsev, the Chief Designer of M4 insisted on its substantial updating in view of the fact that the progress of electronics in 1958-1962 would enable dramatic enhancement of the computer performance, thus making it one order of magnitude more powerful than the then-manufactured Soviet computers. The modernized M4-2M computer had speed of 220 thousand operations per second on programs stored in the read-only memory, main memory up to 16Kwords (29 digits), memory of instructions and constants up to 12K words, (29 digits). This variant of M4-2M was commercially produced during fifteen years since 1964. In 1968 peripheral computers M4-3M were designed for it with the aim of inputting and preprocessing the arriving data, storaging, documenting, and outputting information to the external users at simultaneous asynchronous operation of all user's systems and facilities. The complex of M4-2M and M4-3M had speed of about 400 thousand operations per second.

The head of these designs, M.A. Kartsev, made a substantial contribution to the development of the national digital computers and control computers. He generalized his experience in the monograph "Architecture of Digital Computers" published in 1978 as applied to the third-generation computers.

In 1967, M.A. Kartsev proposed a new approach to architecture and structure of the computer systems based on parallel computations. A draft design of the computer system M-9 with power of 1 billion operations per second was designed in INEUM under the leadership of M.A. Kartsev. Operations over a new class of operands—not numbers but functions of one or two variables defined at discrete points—had to be executed in M-9 on the 32 x 32 matrix of computing elements with common instruction flow. The project of M-9 was rich with new ideas, but many of them have not been implemented until now. Later, M.A. Kartsev's team formed the kernel of a research institute of computer systems bearing now his name. This institute developed high-performance computer systems M-10 and M-13 realizing some concepts suggested in the project of M-9.

The general-purpose computer M-5 intended for planning and economical calculations was developed in 1958-1961 in INEUM under the leadership of I.S. Bruk. It was distinguished for the advanced possibilities of multiprogram and multiterminal operation and, being one of the first domestic computers based on the second-generation circuits and to a large extent was an architectural and structural predecessor of the third-generation computers. It deserves noting that during the development work the INEUM designers of M-5 had no information about the existing computers having the same potentialities. Information about the foreign multiprogram computers such as "Atlas," "Gamma-60," and others became known in this country much later when the design of M-5 was completed.

Automation of powerful energetic "boiler-turbine-generator" blocks at the thermal stations was an important line of INEUM's work in the 1960's. The control computers M-7 designed at INEUM were put in operation at a 200 MW block of the Shchekino station in 1966 and in 1969 at the 800 MW block of the Slavyanskaya station. The M-7-based block control systems maintained normal operation of the blocks and optimized their operation by minimizing the fuel consumption and determining the appropriate setups of the controllers, as well as executed complex logical programs of block startup and shutdown, analysis of combinations of the block parameters with the aim of detecting the pre-emergency situations, and display to the operator of the required information. Design and introduction of M-7 were headed by N.N. Lenov and N.V. Pautin who was the INEUM director in 1964-1967.

In 1965 INEUM headed the project of the USSR Ministry of Instrumentation on the development of the Aggregate system of Microelectronic Facilities of Computer Engineering (ASVT-M) intended primarily for process automation in industry and automatic control systems of enterprises. Although at that time production of the domestic integral circuits was at the stage of development and experimental use, the first control computer systems (UVK) of the third generation were developed in INEUM already in 1970. These systems provided the groundwork of the automated systems that were repeatedly designed in the 1970s for automation of supervision in large power distribution systems, process and enterprise control in machine building, metallurgy, and other industries, as well for automation of research and experiments. B.N. Naumov (1927-1988) who headed INEUM in 1967 (elected in 1984 the Member of the USSR Academy of Sciences) was appointed the Chief Designer of ASVT-M.

The philosophy, structure, and principles of unification of the UVK models, assemblies and devices of ASVT were developed in the late 1960' jointly by INEUM (E.N. Filinov) and NIIUVM (V.V. Rezanov).

The two following basic considerations were taken into account at determination of the principles of architecture and structure of the models of computers and UVK included in ASVT:

o the need for a whole number of models for several hierarchical levels of the industrial automatic control systems satisfying the requirements of problems of different classes such as centralized monitoring of the process parameters, local control of individual technological plants and machinery, supervisory control of manufacturing, planning and so on;

o the possibility of designing general-purpose control computers and systems based on circuitry of the second and third generations instead of the numerous dedicated computers that were designed in the 1950s and early 1960s. The premises for formation of the ASVT philosophy were based on the preceding experience gained by NIIUVM at designing the computer systems of SOU-1 family, Institute of Cybernetics of the Ukrainian Academy of Sciences ("Dnepr" computer), INEUM (M-4, M-5, M-7 computers).

4 New Directions

An architecture supporting program compatibility with the models of the uniform system of computers that were produced commercially in the USSR from the early 1970's was selected for the upper-level models (M-2000, M-3000, M-4000/M-4030). Elaboration of the interface of the upper-level ASUP models to the control systems of lower levels that was carried out within the framework of ASVT paved the way for various projects of joint use of the uniform system of computers and SM EVM. At developing the ASVT software, the problem of compatibility of allied architectures at the level of the operating systems was solved. Being application-compatible with the Siemens OS BS-2000 (architecture of the Siemens 4004 family), the DOS ASVT operating system developed in INEUM under the leadership of I.Ya. Landau and V.A. Kozmidiadi ran on the M-4030 hardware that was binary compatible with the IBM 360/370 and the uniform system of computers (ES EVM). The M-4000/M-4030 hardware was developed at INEUM under the guidance of V.G. Zakharov. M-2000 was produced commercially by the Severodonetsk Instrumentation Plant, and M-3000 and M-4030 by the Kiev Plant of control computers.

Two architectural directions were selected for the medium-level ASVT-M models belonging to the class of mini-computers. The first direction was represented by the models M-6000 and M-7000 developed at NIIUVM under the guidance of V.V. Rezanov and V.M. Kostelyanskii and produced by the Severodonetsk Instrumentation Plant and Kiev and Tbilisi Plant of control computers.

The second architectural direction of the medium-level ASVT-M models was represented by M-400, which also belonged to the class of mini-computers. Its architecture had instruction set and addressing mode that were software-compatible with the PDP-11 family of mini-computers of Digital Equipment Corp., which at that time was internationally recognized as a *de facto* standard, as well as the Unibus system interface to which the peripheral controllers (external memory, I/O) were connected and the controller interfacing the host computer to the object communication devices (USO) from the USO M-6000/M-7000 nomenclature and with the M-40 centralized monitoring machines.

5 Use of CAD

The possibility of connecting the controllers of the hardware connected to the instrumentation interfaces to the Unibus enabled one to design M-400-based problem-oriented measurement-and-computation systems (IVK) using facilities satisfying the CAMAC (Computer-aided Measurement and Control) international standards and ASET (Aggregate Complex of Electrical Measuring Instrumentation). This made

ASVT-M applicable not only in the field of industrial automation, but also in automation of research and experimentation.

Computer-aided design (CAD) became a pivotal application of M-400 which underlay the problem-oriented systems in the form of automated workplaces (ARM) for the radio-electronic and machine building CAD systems including the necessary graphic devices of information I/O and the dedicated software. The Interdepartmental Program for CAD Systems in the Defense Industries headed by the Ministry of the Radio Industry of the USSR and its head organization, the "Almaz" research institute also exemplified formation and implementation of the state policy in science and technology. In these works, the INEUM was represented by I.Ya. Landau and E.N. Filinov. M-400 was commercially produced since 1974 by the Kiev Computer Plant, IVK was produced by the Vilnius Plant of Electrical Instrumentation, and ARM, by the Gomel Plant of Radiotechnical Equipment (GZRTO).

Development in the first half of the 1970s within the framework of ASVT-M of the efficient interactive graphical displays oriented to the architecture of the Unibus computers enabled use of the INEUM achievements in the CAD systems. The characteristics of the M-400-based graphic system and display graphic console EPG-400 (V.I. Fuks) allowed the Central Design Bureau "Almaz" (Academician B.V. Bunkin) to use it as the main platform of the automated designer workplaces for the defense-complex enterprises. Integration of ARM's based on M-400 and EPG-400 was done at GZRTO, which enabled fast introduction of ARM's in the practice of design.

A special place within ASVT-M was occupied by the model M-5000, which also was classified as mini-computers. It was intended to replace the obsolete punch card machines at the mechanized accounting offices of the USSR Central Statistical Department. The original architecture of M-5000 allowed its specialized application to accounting and statistical tasks. M-5000 was designed and commercially produced by the Vilnius Plant of Accounting Machines (Chief Designer A.M. Nemeikshis).

The centralized monitoring and control computer M-40 occupying the lower level in the hierarchy of the ASVT-M models was intended for acquisition, preprocessing, and logging of the process parameters, multichannel two-position control and information output to the digital indicators and CRT's of the operator consoles. M-40 had 1688 (1000 analog and 688 digital) input channels, sensor interrogation speed of 400 channels per second, error of measurement of analog signals of 0.4%. The number of two-positional control channels was 960. M-40 used the microprogram principle of execution of the programs stored for higher reliability in a 16 Kbyte read-only memory. M-40 was designed at INEUM (E.V. Keshek, N.D. Kabanov) and produced commercially by the Moscow "Energopribor" plant (S.Ya. Lebedinskii, V.P. Fedorin).

6 Birth of SM EVM

By a decision of the Intergovernmental Commission on Cooperation of the Socialist Countries on Computers (MPK for VT), in 1974 INEUM was nominated the head organization for design to the small-computer system (SM EVM) and its director B.N. Naumov was appointed the Chief Designer of SM EVM. Since 1984,

N.L. Prokhorov became the director of INEUM and Chief Designer of SM EVM. The complex of SM EVM R&D was carried out by more than thirty institutes and enterprises of the USSR, Bulgaria, Hungary, GDR, Cuba, Poland, Rumania, and Czechoslovakia.

SM EVM included a set of the basic models of micro- and minicomputers such as the basic series of processors of various performance and main memories, wide nomenclature of the I/O facilities, external memory, information display, interface to the object, intra-machine and inter-machine communication. SM EVM was intended for control computer complexes used in the control systems of industrial processes and installations, measurement and computation complexes included in the CAD systems, complexes of data acquisition and processing in the industrial control systems, as well as for small-size commercial and engineering calculations. Since the mid-1970s two mutually complementing international systems of uniform computers and SM EVM became the technical groundwork of automation of control and information processing in all national industries of the participants of the agreement on cooperation in computer engineering.

The advanced technological principles and standards of SM EVM encompassed all aspects of unification of elements, assemblies and devices, designs, models, computers, software with allowance for the technology and power of the national industry and enabled organization of large-scale production on the basis of cooperation of specialized enterprises of various countries.

The design of SM EVM relied on general principles, some of which deserve special mentioning:

o provision of continuity with the previous ASVT-M computers and models:M-400 (SM 3, SM 4, SM 1300, SM 1420), M 5000 (SM 1600),M 6000/7000 (SM -1, SM -2, SM 1210, SM 1634), "MIR" (SM 1410);

o design of systems with task sharing on the basis of general-purpose and dedicated processors of SM EVM;

o wide use of the microprogram control for implementation of the main functions of processors and controllers;

o use of the programmable controllers of the peripherals;

o common nomenclature of the peripherals for some models thanks to the standard interfaces of the peripherals;

o advanced nomenclature of the data-transmission adapters for interfacing the SM EVM computers with the communication lines in compliance with the international standards;

o interfaces of SM EVM to the uniform system computers in heterogeneous systems and networks;

o design of problem-oriented systems produced on the basis of the SM EVM models: measurement and computation systems (IVK) with the CAMAC or ASET GSP hardware, automated workplaces (ARM) for the CAD systems in machine building, radioelectronics, construction engineering, and so on;

o unique form-factors for all SM EVM facilities in compliance with the standards of the International Electrotechnical Commission.

The SM EVM-based IVK of CAMAC or ASET were oriented to real-time automation of complex experiments in various fields of science and technology. Flexibility and modularity of the SM EVM facilities, advanced facilities for interfacing the computer and experiment, problem-oriented system and applied SM EVM software enabled wide use of IVK in the systems of research automation, primarily at the institutions of the USSR Academy of Sciences.

The advent of SM EVM enabled a fundamental enhancement of efficiency and generality of the use of automated workplaces in CAD systems. The potentialities of the universal basic graphical and applied software of the database control systems made real the interactive design, convenient representation of the results, input, editing and output of graphics, diagrams, and drawing. The CAD systems included graphic periphery developed by the defense enterprises for radioelectronic applications (ARM-R), machine building (ARM-M), construction (ARM-S), economics (ARM-E), and so on. Realization of the principle of firmware compatibility of all SM EVM facilities enabled easy successive improvement of the ARM performance by adding the SM 3, SM 4, SM 1420, SM 1700 processors and the graphical vector and raster displays EPG-SM and EPG-3 developed at the institute. (V.I. Fuks).

7 SM EVM Architectures

The design of SM EVM followed two architectural lines. The first line included a wide nomenclature of the control computer systems based on the micro-computers of the SM 1800 family constructed using the bus-model principle (N.L. Prokhorov, A.N. Shkamarda, N.D. Kabanov, A.Ya. Sokolov).

The first models of this line represented 8-bit microcomputers (microprocessor KR580) relying on the bus-model principle with the internal interface I41 (Multibus). In 1986 the first 16-bit model SM 1810 (microprocessor K 1810) of this family was designed, and its manufacture started. All in all, six general-purpose modifications of the SM 1810 and four modifications adjusted to the industrial environment (SM 1814) were developed.

Development of the 32-bit computer system SM 1820 based on the Intel 80386 microprocessor was completed in 1990. All in all, 26 modifications of the SM 1800 family were designed and produced commercially.

A wide nomenclature of external devices, intercouplers to the object, network facilities, adapters and various interfaces (C2, RS422, ILPS, BITBUS, IRPR, and so on) was developed within the framework of this line.

All developments of the SM 1800 family accepted and realized the bus-modular architecture, which enabled virtually continuous evolution of all models of the family both in terms of enhanced performance and satisfaction of the functional requirements of diverse applications.

The system software of SM 1800 included instrumental, real-time executive, and general-purpose operating systems.

Production Associations "Elektronmash" (Kiev) and K.N. Rudnev Plant of control computers (Orel) were the subcontractors at all stages of the development of the SM 1800 family.

The second architectural line of SM EVM was represented by a number of software-compatible models of mini-computer of various performances. The minor models of this line included the 16-bit Unibus-based computers SM 3, SM 4, SM 1300, and SM 1420 (INEUM - B.N. Naumov, A.N. Kabalevskii, V.P. Semik, Yu.N. Glukhov, E.N. Filinov; "Elektronmash" - V.A. Afanas'ev, S.S. Zabara, V.G. Mel'nichenko, V.N. Kharitonov).

SM 1420 then was elaborated to the SM 1425 computer system (N.L. Prokhorov, L.M. Plakhov, G.A. Egorov) using a 22-bit parallel bus system interface featuring more advanced architectural potentialities.

A special place in this architectural line was occupied by the 32-bit mini-computers of the SM 1700 with the Unibus interface and SM 1702 with MPI interface (Chief Designer N.L. Prokhorov, V.V. Rodionov, V.I. Frolov, G.A. Egorov, L.M. Plakhov). The architecture of this family supported virtual memory, software and hardware compatibility with the 16-bit models of mini-computers, as well as an advanced diagnostic system.

The software of this line was represented by a wide spectrum of operating systems, network software for local-area and distributed networks, packages of diverse applied programs.

At developing the SM EVM architecture, original principles of systems with function sharing were elaborated. Owing to that the designers using the circuitry succeeded in implementing two-processor computer systems that were software-compatible with the earlier "MIR" computers for engineering calculations and computers of the M 5000 line for commercial applications.

Controllers and peripheral devices, as well as specially designed processors enhancing substantially the throughput on a particular class of problems occupy an important place in the nomenclature of SM EVM. First of all, the specially designed processor for fast Fourier transforms developed in collaboration with the Institute of Radiotechnics and Electronics of the USSR Academy of Sciences for processing the radar images of the Venus surface (B.Ya. Fel'dman) deserves mentioning. This large-scale study carried out by the USSR Academy of Sciences under the guidance of Academician V.A. Kotel'nikov required computer power equivalent to super-computers. However, the problem was resolved using a mini-computer equipped with a specially designed Fourier processor.

The logic modeling processor representing a dedicated computer for rapid modeling of digital circuits (B.G. Sergeev) deserves special mentioning. Its domain of application is the CAD systems for VLSI circuit design. Original data-flow (pipeline) architecture of the specially designed processor enabled approximately a thousand-fold acceleration of modeling as compared with the general-purpose computers.

8 Conclusion

The aforementioned data about the SM EVM family show that it did not copy the foreign prototypes, relied on the domestic circuits, and was software-compatible with the family of mini-computers that were then most popular in the West. This aim was well justified because otherwise the national computer engineering would be isolated from the international progress in this area and, in particular, would be unable to get access to the accumulated internationally software.

Over sixty thousand computers and control systems of SM EVM, as well as measurement and computer systems (IVK), and automated workplaces (ARM) based on SM EVM were produced from 1974 to 1990 using the INEUM developments.

Importantly, the SM EVM industry had a nationwide infrastructure for technical servicing and training. The SM EVM facilities were a popular school for tens of thousands of experts that were then entering the world of the computer technology.

Ternary Computers: The Setun and the Setun 70

Nikolay Petrovich Brusentsov and José Ramil Alvarez

Moscow State University
ramil@cs.msu.su

Abstract. This paper presents a short history of the development ternary computers "Setun" and "Setun 70" at Moscow State University. It explains the characteristics of their architectures and programming systems (Interpretive Programs), the production and practical usage of Setun-computers, the structured programming on two-stack "Setun 70", the dialogue system of structured programming – DSSP, and the computer assisted teaching system "Nastavnik".

Keywords: Ternary computer, trit, tryte, Setun, DSSP, Nastavnik.

1 Introduction

At the beginning of 1956, the department of electronics came into being at the computer center of Moscow State University (MSU) through the initiative of S.L. Sobolev, a member of the Russian Academy of Science. At the time, Sobolev was the head of the department of calculating mathematics at the mechanical and mathematical faculty of MSU. He initiated a seminar with the aim of providing practical examples for a digital machine usable in universities, in laboratories, and in the design burros of industrial enterprises. It was necessary to create a small computer that was simple to use and master, that was reliable and inexpensive but at the same time was effective in solving a wide range of problems.

A detailed study had begun to determine the status of current computers and their technical abilities. This effort lased a year. The study led to a nonstandard decision to use in the created machine not the binary code but the ternary code. Twenty years later, Donald E. Knuth called this the "balanced" system of calculating and which "may be the most graceful" [10]. Claude E. Shannon revealed some of the advantages of this system in 1950 [12].

The common binary code used in modern computers has the digit set $\{0, 1\}$. Binary code does not contain arithmetical full value, as it is impossible to represent the negative numbers in it. However, a ternary code with a digit set $\{-1, 0, 1\}$ provides optimal arithmetical values for numbers formation. Therefore, we not only an artificial and imperfect complement code of numbers, but we also obtain considerable arithmetical advantages. These include a uniform number code, a variable length of operands, a singular shift operation, a tri-value of a function's "sign of a number", an optimal rounding by simply cutting off lower digits, and a compensation of errors rounding in the process of calculating [7].

J. Impagliazzo and E. Proydakov (Eds.): SoRuCom 2006, IFIP AICT 357, pp. 74–80, 2011.

2 The Setun Computer

To address this new challenge, we had to create a new computer—a ternary computer—called the "Setun" as an experimental model [1, 2]. By the end of 1958, we assembled and began to use the new machine with the collaborators of the electronics department. As we mastered the machine and gained more experience with the program facilities and the variable practical devices, it was obvious that the Setun computer met all requirements and all foreseen tasks intended by its creation.

This success showed the efficacy of the ternary digital machine, even taking into account that the creation of such equipment was the first of such machines completed by a small group of the beginners (eight graduate pupils from MEI and MSU, and twelve technicians and laboratory assistants) in short period of time. In comparison with the binary elements of memory and elementary operations, the ternary design made the architecture of computer more natural and simple.

2.1 Features of the Setun

With a minimum set of commands (only 24 single-address commands), the "Setun" provided an opportunity to do calculations with fixed-point and floating-point numbers. The machine had an index register, the value of which could modify an address by addition or subtraction. It also provided the addition operation with products that optimized polynomial calculations. It utilized three-valued (trit) operations for multiplying and three commands for conditional transition according to the sign of the result. The simple and effective architecture provided an opportunity to facilitate the computer with the programming system and a set of application programs [13] by a small group of programmers by the end of 1959. That was sufficient for interdepartmental testing of the experimental model by April of 1960.

According to the results of testing, the "Setun" computer became "the first working sample of a universal computer on the base of elements without lamps". The machine had the following characteristics: "high productivity, sufficient reliability, small forms, simple technical service". The Council of Ministers of the USSR made a resolution of the serial production of the "Setun" at the Kazan plant of mathematical computers by the recommendation of an interdepartmental committee of experts.

2.2 Production Dynamics

However, for reasons unknown, the officials from Radio Electronically Department did not like the ternary computer and they did not provide the serial production of the computer sample. When it received the opportunity to use the M-20 computer produced in the plant, it did not help the output progress. Notwithstanding, when the number of orders (particularly from abroad) for the Setun grew, they limited the output, which resulted in declining orders and by 1965, they stopped production. In addition, they prevented the mastering of the machine in the USSR where they planned a large production of the machine. The reason for such strange "politics" could have been the very low price of "Setun". It sold for only 27,500 rubles because of the perfect production of its magnetic digital elements at the Astrakhan plants EA and EP where it cost only 3 rubles and 50 kopeks for each element. (The machine had about 2 thousand elements.)

It was essential that electromagnetic elements of the "Setun" accomplished the threshold realization of ternary logic that was truly economical, natural, and reliable. They destructed an experimental model that was in an absolute working state after seventeen years of continuous operation in the computer center of MSU. We changed only three elements with defects during the first year of operation and the machine did not need any repairs to the inner devices. The serial machines functioned well in different climatic zones from Odessa and Ashkhabad to Yakutsk and Krasnoyarsk without any service or spare parts.

Thanks to its simplicity, its natural architecture, and its rational constructed programming system, the Setun effectively used an interpretive system successfully. Some of its features included floating-point numbers with eight decimal digits (IP-2), floating-point numbers with six decimal digits (IP-3), complex numbers with eight decimal digits (IP-4), floating-point numbers with twelve decimal digits (IP-5), auto code "Poliz" with its operating system, and a library of standard programs that used floating-point numbers with six decimal digits.

Users at universities, industry enterprises, and the Scientific Researching Institute (SRI) well mastered the Setun computer. It became an effective tool for solving practical important problems in very different fields from scientific researching modeling and design calculating to weather forecasting and the optimization of enterprise management [3]. Users conducted many seminars on the Setun computer at MSU (1965), at the Ludinovskij locomotive building plant (1968), and at the Irkutsk Polytechnic Institute (1969). Dozens of reports emerged about the economic application of these computers. Thanks to its natural ternary symmetric code, the Setun appeared to be a real universal calculating device. It positively proved to be a technical research tool for computer mathematics in more than thirty universities. For the first time an automated system of computer-assisted instruction occurred at military air engineering academies base on the Setun [11].

3 Some Mathematical Details

The ternary system of calculating applies the same principle of number coding as applied in the binary system used in modern computers. However, the weight of the i-th position is not 2^i, but 3^i. The positions themselves are not two-valued (bits), but three-valued (trits); that is, in addition to 0 and 1, there is a third value, which is -1, thanks to the uniformly represented by positive and negative numbers. We can calculate the value of an n-trit number similar to calculating the value of an n-bit number. That is,

$$N = \sum_{i=0}^{n-1} \alpha_i 3^i = \alpha_{n-1} 3^{n-1} + \alpha_{n-2} 3^{n-2} + \alpha_i 3^i + ... + \alpha_1 3 + \alpha_0$$

where $\alpha_i \in \{-1, 0, 1\}$ is a place value in the i-th position.

The symmetrical system of ternary numbers lends itself to use signs for place value coefficients; that is, instead of using 1, 0, -1, we can use +, 0, $-$. For example, the decimal numbers 13, 7, 6, -6 in such a ternary record would have a representation by:

- o + + + for +13,
- o + − + for +7,

○   + − 0 for +6, and

○   − + 0 for −6.

The sign change of a number using the symmetrical code is equal to a ternary inversion; that is, interchange all "+" values into "−" values and all "−" values into "+" values. Using the above, the representation for −7 would be simply − + −. The definition of addition and multiplying in the ternary symmetrical code appear in the tables below.

Add	+	0	−
+	+ −	+	0
0	+	0	−
−	0	−	− +

Multiply	+	0	−
+	+	0	−
0	0	0	0
−	−	0	+

Unlike the binary this is the ternary arithmetic of signed numbers and the sign of a number is the value of his most significant digit. Problems arising from signed numbers do not have a "perfect solution" in binary code. In the ternary symmetrical code, no such problem exists, which is its advantage.

We can characterize the Setun computer as having a single sequential address, with a 9-trit command code, an 18-trit register summarizer S and multiplier R, a 5-trit index register for address modification F, a pointer for fulfilling commands C, and a one-trit pointer of a result sign ω that can regulate conventional transitions.

The operating memory contained 162, 9-trit cells divided into 3 pages with 54 cells per page for exchange with memory, and a magnetic drum having a size of 36 or 72 pages. Reading and recording into the operating memory occurs using either 18-trit or 9-trit words; a 9-trit word corresponds to the older half of an 18-trit word appearing in the registers S and R. We interpret the content of these registers as fixed number after the second from older position's point; that is, according to module it is less than 4, 5. When calculating with a floating-point mantissa M of a normalized number, it must satisfy the condition that $0.5 < |M| < 1.5$; the order appears as a separate 5-trit word interpreted as whole number with a sign. Page two-stage structure of a memory with word address in the range of three pages that has a 5-trit addresses and a 9-trit commands gave a unusual compact programs and at the same time high speed of the computer in spite of the fact that in interpreting systems magnet drum functions as an operating memory.

4 The Setun 70 Computer

Between 1967 and 1969, we created a ternary digital Setun 70 computer based on the experience and practical application of the Setun computer. An experimental model appeared in April of 1970. This computer contained a nontraditional two-stack architecture designed on the premise of providing good conditions for further expansion and development of its abilities by using an interpretive system method [4].

The Setun 70 used an arithmetic stack (a stack of 18-trit operands) because of the Polish inversion writing program ("Poliz") used as a computer language; Poliz worked well on the Setun computer. The Poliz-program did not consist of commands of different addresses; instead, it used short word logic of 6-trit trytes. As an element of a program, we defined a tryte (the smallest addressable unit) for addressing or for operating. We used an addressing tryte as an operand or as an instruction to send an addressing word (one to three trytes) from the operating memory onto the operand stack. There were only nine pages with 81 trytes per page in the operating memory; for the access, we opened only three pages with their numbers in the "pointing registers". The operating tryte points the operations or procedures for the operands stack and the processor register. There are only 81 operations: 27 basic types, 27 auxiliary types, and 27 that are programmable by a user.

The second system stack had return addresses when processing breaks. The fulfillment of the program allowed the realization of E.W. Dijksta's idea of structured programming on the Setun 70. This allowed for an input call operation of a program, a call according to a condition, and a fulfillment of a program cycle. As a result, the machine confirmed that procedural structured programming highlighted the advantages of Dijksta's method; that is, the difficulty of creating programs became five to seven times less, thanks to the traditional testing "arrangement" on examples we used. Moreover, the programs became more reliable, logical, understandable, and easily modifiable. In addition, the peculiarities of the Setun 70 became a basis for dialogue system of the structured programming DSSP that was accomplished on the DVK type computers and on personal computers [6, 8].

5   The Setun 70 and the "Nastavnik"

Unfortunately, administrative decisions stopped the further development of the abilities of the Setun 70 computer and its programming capabilities. We had to change our focus toward the study of computerization. The Setun 70 became the basis for development and fulfillment of automated computer systems for the teaching of "Nastavnik" [5, 9] that contained the principles of Y.A. Kamenskij's "Great Didactics". The purpose of the computer in this system was not the "electron turning over pages" and multimedia effects; instead, it checked the students' understanding of what they were studying in an attempt to overcome misunderstandings and providing with real subject skills by exercises. In addition, the computer would record the lesson itself giving the creator the opportunity to evaluate the effectiveness of using didactics methods and to modify the material and improve them.

Studying materials in "Nastavnik" appeared in typed form with numbered sections, passages, exercises, and notice of wrong answers. Thanks to the Setun 70, the computer was able to interact with a student rather easily. It gave the "book" the ability of communicating with a reader with the help of simple terminal with digital keyboard, and calculating indicator. The creation of study materials for the "Nastavnik" was not associated with computer programming; in practice, it showed that schoolteachers could create materials in mathematics, physics, English, and other subjects. The didactic effectiveness of the system was very high. For example, for the course in "Basic Fortran", students of MSU studied on the "Nastavnik" for ten to fifteen hours, students from the economics department for fifteen to twenty hours. The results showed that students using the computer had better programming skills in Fortran than those studying a usual semester course on the subject.

The principle of using a "book-computer" in the "Nastavnik" provided an opportunity for optimal computer usage as a practical means for studying. In all respects, the Setun computer provided the necessary tool (microcomputers and connected with it, 30-40 terminals as a simple calculator) for automated learning; it was inexpensive, reliable, and easily mastered by students and teachers. The work in a regime of a dialogue was not difficult; in fact, it was fascinating and guaranteed quick and good mastering of a subject that had good material organization. The application system at MSU, MAI, VIA, in schools, and for professional studying at a plant proved its effectiveness in wide range of subjects and study levels. The "Nastavnik" functioned for thirty years for automatic testing and tests to identify the English language level of first year students to form groups according to levels.

Despite the great necessity to improve the process of studying in the "informative" twentieth century, the "Nastavnik" began to fall in disfavor. The reasons may include its low price, its simplicity, and the lack of a proper display, mouse, and hypertext. The fact remains that using information technologies in the process of studying elevates student learning according to the number of computers used and their power and not to their level and quality of teaching.

References

[1] Brusentsov, N.P.: Computer 'Setun' of MSU. New Developments in the Field of Calculating Mathematics and Computers, 226–234 (1960)

[2] Brusentsov, N.P., Maslov, S.P., Rozin, V.P., Tishulina, A.M.: A small computer 'Setun', MSU, p. 145 (1965)

[3] Brusentsov, N.P., Morozov, V.A.: Annotated guide of programs for Setun, MSU, #2 (1968/1971)

[4] Brusentsov N.P., Zhogolev, E.A.: Structure and algorithms of small computer functioning. Computers and Problems of Cybernetics #8, 34–51 (1971)

[5] Brusentsov, N.P., Maslov, S.P., Ramil Alvarez, J.: Automated system teaching 'Nastavnik'. Computers and Problems of Cybernetics #13, 3–17 (1977)

[6] Brusentsov, N.P., Zlatkus, G.V., Rudnev, I.A.: Dialogue system of structured programming. Programming Rigging of Microcomputers, 11–40 (1982)

[7] Brusentsov, N.P.: Notes about ternary digital equipment. Architecture and Programming of Digital Systems, 114–123 (1984)

[8] Brusentsov, N.P., Zakharov, V.B., Rudnev, I.A., Sidorov, S.A., Chanishev, N.A.: Developing adapting language of a dialogue system of programming, MSU, p. 80 (1987)

[9] Brusentsov, N.P., Maslov, S.P., Ramil Alvarez, J.: Microcomputer system teaching 'Nastavnik', Nauka, p. 223 (1990)

[10] Knuth, D.E.: Seminumerical algorithms. The Art of Computer Programming, vol. 2. Addison-Wesley, Reading (1969)
[11] Kuznezov, S.I.: Materials on mathematical service of "Setun" (1964)
[12] Shannon, C.E.: Symmetrical notation for numbers. The American Mathematical Monthly 57(2), 90–93 (1950)
[13] Zhogolev, E.A.: The Order Code and An Interpretative System for the SETUN Computer. USSR Comp. Math. and Math. Physics #3, 563–578 (1962)

Ternary Dialectical Informatics

Nikolay Petrovich Brusentsov and Julia Sergeevna Vladimirova

Moscow State University
{ramil,vladimirova}@cs.msu.su

Abstract. Modern binary informatics with its so-called "classical" two-valued logic admits to create an artificial intellect and suppresses the natural intellect of students and other thinking people. Logic that based on dogmatic law of the excluded middle is incompatible with dialectical principle of opposition coexistence. Such logic is deprived of fundamental logical relation – the content consequence, and then cannot reach a conclusion. Aristotle's syllogistics includes the content consequence as common affirmative premise "All x are y". However, binarity misinterprets it as a paradoxical material implication that is not a relation at all. Lewis Carroll's "Symbolic logic" correctly represented syllogistics, but it is not intelligible for modern binary logic. Our paper reveals the essence of content consequence; it explains Carroll's intentional judgments and syllogistic relations submitted to opposite coexistence.

Keywords: Ternary informatics, opposition coexistence, content consequence, Aristotle's syllogistics, Carroll's "Symbolic Logic".

1 Introduction

The deficiency of modern binary informatics is specified with inadequacy of its foundation is adequate for its specification since it uses binary logic. This so-called "classical" formal logic, based on prior transcendental law of the excluded middle, does not show the perfect reflection of the reality and it is out of keeping with common sense. It is so opposite to practical arithmetic, for example, it is not useful for solving real problems. This binary logic is the logic of an artificial, discrete world of binary computers. There is no modality; possibility cannot differentiate from necessity. Even the most fundamental logical relation of content consequence has expressions by "material implication". These paradoxes for which distinguished logicians have tried for many years to overcome in vain.

It is clear why logic cannot become a school subject. The development of logical thinking or the study of logic, first as school task and then as a university one, does not lead to mentality improvement. The situation became worse with computerization of the education. With the general discussions concerning the intellectualization of computer processing of information and information safety, there is a suppression of people's intellect everywhere with stilted binary logic.

Most agree that Aristotle is the founder of logic. He created a system of demonstrative conclusions – the syllogistics that are still an unsurpassed mentality

J. Impagliazzo and E. Proydakov (Eds.): SoRuCom 2006, IFIP AICT 357, pp. 81–88, 2011.
© IFIP International Federation for Information Processing 2011

instrument. Syllogistics is dialectic; there is no paradox in it, but we cannot transform it into a modern logical calculus. This motivated suspicions that there was something wrong in Aristotle's logic such as his denial of empty sets. However, there are not only empty sets in syllogistics but also fuzzy sets that were invented by L. Zadeh in 1965 and these are still not "mastered" with modern logic. The principal difference of Aristotle's logic from modern, "classical" one is that it is not binary; rather it is ternary. Contrary to the "law of excluded middle" with "necessary is" and "necessary isn't", there is the third one affirming, "possible is and possible isn't". We predicate the "threevaluedness" to the relation of consequence that depletes the determined by Aristotle in the "Prior Analytics" [paragraph 57 line number l]:

> "...when two things are so related to one another, that if the one is, the other necessarily is, then if the latter is not, the former will not be either, but if the latter is, it is not necessary that the former should be. But it is impossible that the same thing should be necessitated by the being and by the not-being of the same thing"

2   Mathematical Considerations

In syllogistics, we present the consequence relation as universal affirmatives with premise "any x is y". In this case, we see that any x-thing is necessarily a xy-thing, and any y'-thing (not y-thing) is necessarily an $x'y'$-thing. Besides, a xy'-thing must be an exclusion as x must necessarily be y (x cannot be y'). At the same time, we cannot exclude $x'y$-things; they are possible but not necessarily, as there are xy-things and $x'y'$-things. If we assume both premises,—"any x is y" and "any y is x"—then we must exclude xy'- and $x'y$-things. Hence, there will exist an equivalence relation—"x is interchangeable to y". In binary logic, we would have to admit "interchangeable/not interchangeable".

For the relation of consequence, we need three values, which are "necessary is", "possible but not necessary", and "impossible". Therefore, there is not enough binary implication for the adequate expression of consequence. With xy'-things, consequence is impossible; if there are no xy'-things, then we cannot exclude consequence; otherwise, it is possible but not necessary. In case of inexistence of x-things or y'-things, the implication does not express any interconnection between terms and does not form a two-dimensional relation. In these cases, the other term can have any value independently.

To smooth away "paradoxes" of implication, it is enough to prevent these cases of the invariability of its terms. Hence, a strict implication of Lewis is an inexistence of xy'-things; that is, $V'xy'$ is paradoxically performed when x-things do not exist and y'-things do not exist—or when $V'x$ and $V'y'$. There would not be paradoxes if with $V'xy'$ we demand Vx and Vy'—the existence of x-things and y'-things. As the result, the implication of Lewis becomes the necessary consequence $VxV'xy'Vy'$ [3]. In full normal form of this relation, $VxyV'xy'Vx'y'$, is visibly ternary. Members xy and $x'y'$ belong to concerned subset Cartesian product $\{x, x'\}x\{y, y'\}$, member xy' antibelongs to it and member $x'y$ is withheld. Withholding expresses the third variety of belonging that is possible, but not necessary. A subset that admits a possible belonging is fuzzy and a relation presented with it is ternary.

In the mathematical logic, deviation from Aristotle's opinion of the universal affirmative premise that "All A is B", takes it from a content consequence to binary implication. Hilbert and Ackerman proved the deviation with mathematical application logic needs "where taking Aristotle's opinion as a foundation was pointless" [7, p.79]. They did not take into consideration that the logic lost the richness of content, thinking that their logical calculus "makes possible successful problem comprehension where simple content logical thinking is principally weak" [7, p.17].

Indeed, logic became mindless two or three thousand years ago. Ancient stoics sought after incredible abstraction and carried it out with the help of "propositions" subordinated to "law of excluded middle". That law allowed only two truth-values— "true" and "false". Adequate to reality, Aristotle's syllogistics became a dead scholasticism with binary logic. Mathematical logic presented this "classic" thinking with strict algebraic forms evidently showed its inadequateness [8].

Stoics "compensated" for the lack of the consequence relation in their logic to implement conclusions according to the rules of *modus ponens* and *modus tollens*. In mathematical logic, we know that an implication is not a consequence. "The relation 'if X then Y' should not be understood as a form for relation of foundation and consequence. On the contrary, the proposition $X \rightarrow Y$ is always true when X is false or when Y is true" [7, p. 20]. At the same time, however, in mathematical logic even its founders identify binary implication with ternary and with Aristotle's consequence as the first and the second one associated with the proposition "All A are B". As a result, from mathematical logic point of view, they consider the perfect forms of syllogisms *darapti, bamalip, felapton, fesapo* to be false [7, p.79]. The syllogism of submission of quotient to common is rejected as according to the deviation from Aristotle's interpretation "Some A is B" is not a consequence from "All A are B".

Yan Lukasevitch created a ternary modal logic in 1920 in his detailed book titled, "Aristotle's syllogistics from a modern formal logic point of view" [9]. He proved algebraically with the help of the identification of the ternary consequence the material implication of Aristotle's statement— "But it is impossible that the same thing should be necessitated by the being and by the not-being of the same thing" —is false. It is also false from the logical point of view that they did not obey the basic logical law – the law of identity. We cannot identify the proposition "All A are B" that expresses the consequence relation B from A with "No one A isn't not-B", the help of which binary implication relation is presented in natural language.

The problem is that logic cannot be without a consequence relation that is in essence is ternary; furthermore, it does not exist in binary logic. We express the relation called implication like consequence with the same "If... then..." relation and the same "$\rightarrow$" marker. It is not striking that we can interpret implication as consequence. However, if there is no logic without consequence then logic with implication instead of consequence is not logic at all! Then from inexistent things follows "whatever", from 2x2=4 is that "snow is white".

Defective binary logic ignores common sense. The result of its application does not meet with natural expectations. In the book by T. Oppenheimer [10], he proves the indisputable bad influence of computer education in American schools. The author insists on removing computers from schools; but it is hardly possible in current situation. The "root of all evil" is not with computers but with the primitive unnatural logic of the discrete binary world that students study. It blocks their ability to master the logic of real world. If there were natural logic in computers, the result of studying computers would be the opposite.

However, there is no adequate (dialectical) logic in all "science of thinking", even where it does not follow the "law of excluded middle" and focuses on invention of not binary logics. This invention is unsuccessful because it has a formal character. If they researched the essentials of the problem, then they would have found out that Aristotle's logic is ternary and that ternary is a necessary, but not a sufficient, condition for adequate logic. Aristotle's logic is adequate; hence, it makes no sense to invent non-Aristotle logics.

3 Carroll Logic

A good exception is "Symbolic Logic" by Lewis Carroll [6] that did not receive appropriate (like Aristotle's one) understanding and development. There is neither discontent "true" or "false" propositions nor "law of excluded middle". His logic researches propositions that expressed interconnections of things characterized with combinations of features (peculiarities).

> "The Universe contains 'Things' ... Things have 'Attributes' ... Any Attribute, or any Set of Attributes, may be called an 'Adjunct' "

The proposition is considered as a natural language expression of a relation that connects adjuncts of things with terms x, y, z, ... At the same time, the essence of Carroll's relation visually reflects his diagram and algebraic "index method" that formally allows obtaining a content conclusion from data opinions, if they exist.

Carroll's diagram is on the surface identified to Pirs' truth table used to identify Boolean functions. However, expressed with the diagram, we interpret it not as extensional (as class of things) but intentional as set of things or subset of Cartesian product of pairs of opposite adjuncts. Besides, the diagram cells obtain not one from two values, but one from three values; it could also obtain cells that have "0" or "1" and they permit empty cells. They denote nothingness of the belonging of things to the subset presented in the diagram. Nevertheless, Carroll understood value "1" as existence, value "0" as inexistence of a thing; empty cells belong to neither.

For example, Lewis' strict implication relation $V'xy'$ at Carroll's two-term diagram only one value appears— "0" in the xy'-cell. Carroll expressed this relation in three ways using a universal negatives proposition: "No one xy' exists" or "no one x is y'" or "No one y' is x". The universal affirmative proposition "All x are y" includes a particular affirmative in Carroll's "Some x is essence y" that is equal to existence of the proposition "Some xy exist". Carroll called "All x are y" a double proposition that was equal to two opinions: "No one x is y'" and "Some x is y"; that is, $V'xy'Vxy$. In the diagram, "0" is in the cell xy' and "1" is in the cell xy.

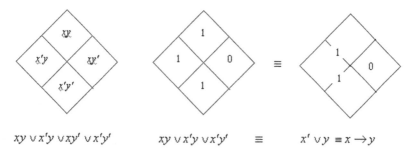

$$xy \vee x'y \vee xy' \vee x'y' \qquad\qquad xy \vee x'y \vee x'y' \qquad \equiv \qquad x' \vee y \equiv x \rightarrow y$$

Fig. 1. Material Implication: Extensional interpretation (Pirs' table, Boolean class algebra)

4 Mathematical Details

It is clear that this is not implication (one of its paradoxes is eliminated) but it is not a full Aristotle consequence. He does not take into consideration the contraposition of a consequence – a false step, as it is peculiar to all efforts to make syllogism algebraic. As compensation [1], "Symbolic logic" of Carroll becomes a well-composed and perfect statement of Aristotle's syllogistics – a foundation of dialectical logic.

The most important component of logic content appeared an identified dialectic principle in the Aristotle's syllogistics – the principle of opposition coexistence [2]. It expresses the contraposition of a universal affirmative proposition as the symmetry of a relation expressed with a universal negative proposition; it constitutes the visible demonstrations of this principle. The fact is that the initial adjuncts x, x', y, y', z, z' ... expressed with terms x, y, z ... made sense as the result of things comparison that have opposite adjuncts, for example, an x-thing and an x'-thing. In other words, the principle of opposition coexistence means that a subset of the Cartesian product $\{x, x'\} \times \{y, y'\}$ reflects a content relation that has all pairs opposite adjuncts – $VxVx'VyVy'$. In Carroll's diagram, $VxVx'VyVy'$ reflected with token data of "1" existence on every one of four interior fences that denotes non-emptiness of classes x, x', y, y'.

These equivalent realities are part of Aristotle's Universe (AU) – the foundation of content logic [2]. In it Lewis' implication $V'xy'$ and Carroll's $VxV'xy'$ become a full consequence.

$$(V'xy')(VxVx'VyVy') \equiv VxyV'xy'Vx'y'$$
$$(VxV'xy')(VxVx'VyVy') \equiv VxyV'xy'Vx'y$$

Inexistence of any possible in the diagram things (e.g. an xy-thing in AU) means the existence of two adjoining things with it, so:

$$V'xy' \equiv (V'xy)(VxVx'VyVy') \equiv V'xyVxy' \, Vx'y$$

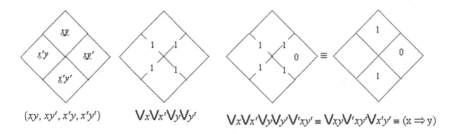

$\{xy, xy', x'y, x'y'\}$ $V_xV_{x'}V_yV_{y'}$ $V_xV_{x'}V_yV_{y'}V'_{xy'} \equiv V_{xy}V'_{xy'}V_{x'y'} \equiv (x \Rightarrow y)$

Fig. 2. Content Consequence: Intentional interpretation (Carroll's diagram in Aristotle's Universe AU)

The existence of a xy'-thing according to the principle of opposition coexistence entails the existence of its antipode – an $x'y$-thing. Therefore, the particular affirmative and the particular negative premises of the syllogism become double and they are two, not four:

$$Ixy \equiv Axy \vee Ayx \equiv VxyVx'y'$$
$$Oxy \equiv Exy \vee Ex'y' \equiv Vxy'Vxy'$$

At the same time, common conditions are not two but four that come from one by term inverting:

$$Exy \equiv Axy', \; Ex'y' \equiv Ax'y, \; Ayx \equiv Ax'y'.$$

Syllogistics algebra that corresponds to Carroll's interpretation of its diagram with token symbolized existence and inexistence of things compared with cells and "fences" in similar to "index method" but instead of indexes it uses prefix functor of existence V – "disjunct" (integral disjunction similar to integral sum Σ) and its inversion V' – symbol of inexistence. The illustration presented in the diagram relation reflects a conjunction of disjuncts, not inverted and inverted, and members of conjunction that corresponds to empty cells that are not present (silent). Every trit takes one of three values: "+" for existence, "–" for inexistence, and "0" for silence. For example, consequence $(x \Rightarrow y) \equiv Vxy'Vx'y'Vx'y'$ is reflected with the value of a four-trit scale: +–0+, a particular negative premise Oxy coded with the value: 0++0. There are eight double place relations in syllogistics [5]:

$$Axy \equiv Ay'x' \equiv Exy' \equiv Ey'x \equiv +-0+$$
$$Ayx \equiv Ax'y' \equiv Eyx' \equiv Ex'y \equiv +0-+$$
$$Exy \equiv Eyx \equiv Axy' \equiv Ayx' \equiv -++0$$
$$Ex'y' \equiv Ey'x' \equiv Ax'y \equiv Ay'x \equiv 0++-$$
$$Ixy \equiv Ix'y' \equiv Iyx \equiv Iy'x' \equiv Oxy' \equiv Oyx' \equiv Ox'y \equiv Oy'x \equiv +00+$$
$$Oxy \equiv Oy'x' \equiv Oyx \equiv Ox'y' \equiv Ixy' \equiv Iy'x \equiv Ix'y \equiv Iyx' \equiv 0++0$$
$$x \Leftrightarrow y \equiv AxyAyx \equiv Exy'Eyx' \equiv +--+$$
$$x \Leftrightarrow y' \equiv ExyEx'y' \equiv Axy'Ayx' \equiv -++-$$

Computerizing proof of conclusions (true modi of syllogism) is carried out by two terms premise presentation with three term scales from witch crossing searching conclusion is taken by middle term elimination, if it exists. For example, modus *Barbara*: $AyzAxy \Rightarrow Axz$ in three term x, y, z-scales is realized:

$$Ayz \equiv +-0++-0+$$
$$Axy \equiv ++--00++$$
$$Ayz \cap Axy \equiv +---000+$$

The elimination y gives x, z-scale +−0+, i.e. Axz. Subordination of particular premises is proved with a common crossing of coding these premises scales. So, the subordination $Axy \Rightarrow Ixy$, is equal to $AxyIxy = Axy$, proved with crossing

$$+-0+ \cap +00+ = +-0+.$$

In syllogistics based on the opposition coexistence of all doubtful modi from the classical logic point of view and a row of modi missed with traditional syllogistics. For example, from premises of doubtful modus bamalip, there is not only particular but also a universal conclusion:

$$Azy \equiv +0-++0-+$$
$$Ayx \equiv ++00--++$$
$$Azy \cap Ayx \equiv +0-0---+$$

Eliminating y we have

$$+0-+ \equiv Azx \equiv Ax'z',$$

i.e., $AzyAyx \Rightarrow Azx$

The correction of traditional theory is proof of denying with it modus of the first figure $IyzAxy \Rightarrow Ixz$:

$$Iyz \equiv +00++00+$$
$$Axy \equiv ++--00++$$
$$Iyz \cap Axy \equiv +0--000+$$

That with y elimination is +00+, i.e. Ixz. By the same way, the true of the next missed modus is proof of the first figure $IyzExy \Rightarrow Oxz$ and similar modi of other figures.

References

[1] Brusentsov, N.P.: Diagrams of L. Carroll and Aristotle syllogistics. Computers and Problems of Cybernetics Questions (13), 164–182 (1977)
[2] Brusentsov, N.P.: Wandering around three pines (dialectics adventures in informatics) (2000), http://ternarycomp.cs.msu.ru/Papers/3PINES.pdf
[3] Brusentsov, N.P.: Ternary interpretation of Aristotle syllogistics. Historical Mathematical Researches 43(8), 317–327 (2003)
[4] Brusentsov, N.P., Vladimirova, J.S.: Ternary computerization of logic. In: 12th Russian Conference on Mathematical Methods of Form Identification, Report Digest, pp. 40–42 (2005)
[5] Brusentsov, N.P.: Reanimation of Aristotle's syllogistics. Logic Restoration, 140–145 (2005)

[6] Carroll, L.: Symbolic logic. L. Carroll Story with Bundles, 189–361 (1973)

[7] Hilbert, D., Ackermann, W.: Foundations of Theoretical Logic (1947)

[8] Losev, A.F.: Critical notes about bourgeois mathematical logic. Historical Mathematical Researches 43(8), 339–401 (2003)

[9] Lukasevitch, Y.: Aristotle's Syllogistics from Modern Formal Logic Point of View (1959)

[10] Oppenheimer, T.: The Flickering Mind: The False Promise of Technology in the Classroom and How Learning Can Be Saved, p. 512. Random House, New York (2003)

Establishing a Computer Industry in the Soviet Socialist Republic of Belarus

Yuri Vladimirovich Karpilovitch[1], Viktor Vladimirovich Przhijalkovskiy[2], and Gennadiy Dmitrievich Smirnov[3]

[1] Minsk Institute of Management
[2,3] Scientific Institute of Digital Electronic Computer Technologies
przy@yandex.ru

Abstract. Created in the late 1950s in Minsk, for several years the factory for the production of computers occupied a leading position in the Former Soviet Union by the number of general-purpose computers produced. For many years, the Minsk Plant at GK Ordzhonikidze produced about seventy percent of all computers manufactured in the Union. This paper highlights the activities undertaken by the leadership of the Belarusian Republic that contributed to this production. It illustrates the basic technical specifications of computers developed by the Minsk computer factory that includes the Minsk-1, the Minsk-2, theMinsk-22, the Minsk-23, and the Minsk-32 machines. This paper also shows the timing of their development, the number of units produced, and the names of the main developers.

Keywords: Minsk computer, NIIEVM, Minsk Factory for Computers.

1 Introduction

In August 1956, after completion of the creation the first computers in the USSR ("Strela", BESM, M-3, "Ural -1"), the Council of Ministers of the USSR issued the regulation about measures for a considerable increase in the manufacturing of computers in the country. This decision provided the building of several factories for the computer, their modules, and other details. The factory of mathematical machines in Minsk would become one such new factory according to the decision.

The management of Sovnarhoz BSSR, understanding the importance of the new enterprise for a development of the city, allocated a building site for it at a not-yet completed garment factory building on one of city main squares – the Jakub Kolas square. This decision speeded up essentially a formation of the new factory, which they named the "Minsk Factory for Computers" of G.K. Ordzhonikidze.

In 1958, the special design bureau (SDB) was organized at the factory under construction for the manufacture and modernization support of producing computers. SDB staffing also accelerated by the republic and city authorities who invited some tens of experts who already had computer development experience from other cities in the USSR. Among those invited were G.P. Lopato from Moscow, V.J. Simhes, A.I. Bahir, G.K. Stoljarov, and A.P. Zhigalov from Zagorsk, E.I. Sakaev, S.N. Remorov,

J. Impagliazzo and E. Proydakov (Eds.): SoRuCom 2006, IFIP AICT 357, pp. 89–97, 2011.

V.V. Przhijalkovsky, V.E. Klochkov, N.A. Maltsev, and R.M. Astsaturov from Noginsk, V.A. Averyanov from Penza, and I.K. Rostovtsev, G.D. Smirnov, and J.G. Bostandzhja from Yerevan. Together with the Minsk experts, they headed the development and computer manufacturing in the SDB and at the factory. It essentially reduced terms of formation of the Minsk industrial complex on computer production. The prestige of the new enterprise and its development control of the authorities attracted attention of the best experts of a city.

The integration within the limits of an enterprise of developing industrial departments played the positive role since the beginning. It allowed combining a process of development of a product with preparation of its serial production that led to essential reduction of terms of producing new computer manufacturing and fast growth of rates of their output.

Resource conflicts between maintenance of current manufacture and development of the next product were resolved by administration (the director was V.K. Goldberg, the chief engineer was N.I. Kiriljuk). In concrete conditions of the Minsk computer factory of G.K. Ordzhonikidze the perspective policy of a management led to that between the development termination (the state tests) and output of the customizing party of computers in some cases (Minsk-1, Minsk-22, Minsk-23, Minsk-32, ES-1020, ES-1022) made terms from one to three months. At that time, the terms of development of the next model were under two years, and the estimate budget of development was record-breaking small.

With growth of complexity of computers, and with the advent of new manufacturers (e.g. the Brest electromechanical factory in 1966), with the expansion of the nomenclature of developed products, and with the increase in expenses at development, it became impossible to maintain the developed organizational unity within the limits of an existing economic mechanism. The SDB factory was allocated for independent balance at first (MPB – 1966) and later it received full economic independence (branch NITSEVT – 1969г., NIIEVM – 1972г.)

2 Before the Minsk Computers

In the spring of 1959, in response to a post of the chief engineer of SDB of factory of G.K. Ordzhonikidze, they invited G.P. Lopato, one of the developers of the M-3 computer. In 1956, he created the Electro-modeling laboratory of the Academia of Science of the USSR together with the All-Union scientific research institute of electro-mechanics of Academia of Science the USSR. The production facilities were ready to generate computer production; only the staff planned and completed the developments in SDB. In these conditions, the decision of administration to accept the M-3 computer as a starting product was natural and reasonable.

In September of 1959, the factory produced the first computer made completely independently. It was the variant of the computer with operative memory on a magnetic drum. Its speed was only thirty operations per second, but on it, many technological processes were formed; specialists of development and service were staffed and trained. In 1960, the development of for M-3 operative memory on ferrite cores in capacity 1024 31-bit words has finished. The work fulfilled by G.P. Lopato, V.J. Simhes, E.I. Sakaev, A.I. Bahir, V.A. Harlap, and V.B. Novysh.

By the end of 1960, they produced twenty-six M-3 computers and from them, ten computers had the ferrite memory, which increased their speed to 1000 operations in one second. Programming M-3 computers was in machine codes in an octal number system. For the Minsk factory, this computer was valuable as it was the starting one. Sadly, its influence on further computer development was almost zero.

3 Birth of the Minsk Machine

In August of 1960, they completed the development of the first independent computer the Minsk-1. Its features included 800 lamps (valves or vacuum tubes), 2500 operations per second, ferrite memory with 1K words, a word length of 31 bits, a two-address instruction set fixed before the high bit, a peripheral memory on the magnetic tape of 64K words, punched tape input of 80 words per second, and output to digital printing of 20 words per second. G.K. Lopato was the chief designer of the computer; the main developers were E.I. Sakaev, V.Y. Simhes, A.I. Bahir, V.L. Salov, S.N. Remorov, G.K. Stolyarov, B.I. Tsagelsky, and G.M. Gendelev.

The Minsk-1 computer had no compatibility with the M-3 computer. It inherited from its predecessor only two architectural characteristics: word length and two-address instructions. The Minsk-1 computer was produced until 1964 and it had only a few completely compatible modifications during its productive life.

The Minsk-11 computer was intended for seismic information processing and operation with remote users. The chief designer was V.M. Manzhalej. They produced eleven computers of this version. The Minsk-12 computer had the expanded RAM in capacity 2048K words and tape drives with a capacity of 100K words. The chief designer was V.Y. Simhes. They produced five versions of this computer.

The Minsk-14 and Minsk-16 computers were intended for telemetry information processing, so they included reading devices of for telemetry. The main designers were L.I. Kabernik and V.M. Manzhalej. In all, they produced thirty-six Minsks-14 computers and one Minsk-16 computer.

In addition, they produced the Minsk-1 computer as a system for storage and recognition of fingerprints for the Ministry of Internal Affairs of the USSR. A.M. Tolmachyov was the main designer of this system. In total between 1960 and 1964, they produced 220 Minsk-1 computers. They, along with "Ural" computers, became the most mass produced computers in the country during that period.

They used machine code to program the Minsk-1 computer. A delivery package included a routine library containing about one hundred programs with 7500 commands. During this period at SDB, they carried out the serious operation for developing the first systems of auto-programming; that is, they developed translators for the autocode "engineer" and the autocode for the "economist". They fulfilled the work at the laboratory of programming of SDB led by G.K. Stoljarov, where they grew a whole group of known programmers such as M.S. Margolin, M.E. Nemenman, E.V. Kovalevich, V.I. Tsagelsky, and N.T. Kushnerev.

4 The Minsk-2 Series of Computers

The next development of SDB of the factory of G.K. Ordzhonikidze was the second-generation Minsk-2 computer. V.V. Przhijalkovsky was the chief designer of this development. The main developers are V.E. Klochkov, G.D. Smirnov, N.A. Maltsev, A.I. Bahir, J.G. Bostandzhjan, V.K. Nadenenko, G.K. Stoljarov, and M.B. Tyomkin.

The computer was under development between1960 and 1962, at the same time with the Minsk-1 production. The situation did not demand to ensure a software compatibility of these computers. In addition, to ensure such compatibility it would be far from simple, at least because of the hopelessness of a 31-bit grid. For the new computer they used the 37-bit grid that contained the number sign and twelve octal or nine decimal bits of number. With the Minsk-2 computer, a first for Minsk computers, they used floating-point for the representation of numbers; in this connection, they employed seven bits for order representation, including the order sign bit. Other bits in this case represented a mantissa of number with its sign. It provided representation of numbers from 10^{-19} to 10^{+19}, which was quite enough for a general-purpose small computer.

For the first time in domestic computer facilities with the Minsk-2 computer, it provided the processing of the alphanumeric information in an explicit form. In a machine word, six alphanumeric characters occupied places, encoded by the telegraphic code MTK-2. In this case, they used the sign digit to divide documents and messages among them.

The command of the Minsk-2 computer consisted of an opcode (seven bits), the number of the block of the RAM (two bits), the address of an index cell (four bits), and two twelve-bit addresses. Thus, it provided the functioning of 127 commands, 15 index registers, and the addressing of 8196 words in RAM.

Moreover, the two-address command system of the computer contained standard arithmetic commands and input-output commands. The command of the special arithmetic unit provided the realization of operations for double precision; it included the original command of cycle and a variety of the commands essentially accelerating information processing. Of the one hundred commands involved in the computer, forty were arithmetic. Twenty-seven commands were not involved with arithmetic in the Minsk-22 computer; part of these involved the extracodes and the rest were in reserve for the organization of specialized systems.

The designers developed for the Minsk-2 computer a system of semiconductor units based on inexpensive and widespread transistors of type P-16A. The Potentially-pulse system of units used diode-transformer circuits on oxy-ferrite cores. The clock rate of the complex achieved 250k Hz.

The units were constructed on replaceable cells having double-sided printed circuit wiring and the printing plugs. Out of twenty-three types of cells, five types made 70% of all the equipment. In total, the computer included 1286 cells that contained 7500 transistors and 18,000 diodes. Thanks to simplicity, cheapness, and reliability, this complex of units was operational for six years in the Minsk-2/22 and Minsk-23 computers.

For the Minsk-2 computer, they developed a ferrite storage device with semiconducting handle. This was perhaps the first semiconductor storage device of size 4096 words in the USSR; A.I. Bahir, J.G. Bostandzhjan, and V.A. Harlap were its designers. The RAM worked on ferrite cores of external diameter of 1.4 mm; it had a cycle of 20 microseconds, a latency of 7.5 microseconds and it contained 740 transistors and 1550 diodes. The power consumed by RAM was only 800 watts. At that time in 1962, it was a pioneering and very successful development. The speed of the computer was 5000-6000 two-address operations per second.

They developed for the computer a new tape drive with a bit density of 12 impulses per millimeter and a photo-reading mechanism for paper punched tape working at a speed 800 strings per second. The listing of the alphanumeric linformation was made on the RTA-50 rolled telegraph.

In the Minsk-2 computer, a first for Minsk computers, they used a hardware-software interruption of programs by a method of operation that suspends the arithmetic device. They used it for operations with output devices of the information and operation with "extracodes". Extracodes, or macrocodes, were also an innovation for Minsk computers. Interruption of programs and extracodes were planned ahead and more effectively used in the expanded package of the Minsk-22 computer.

The complete computer consoles had taken 40 square meters of space and it consumed from a three-phase network of 380/220V less than 4 KW. The development of the Minsk-2 concluded in September of 1962. In 1963, its production began from the factory of G.K. Ordzhonikidze. By the end of 1964, they produced 118 computers.

By the end of 1964, they completed the operations on developing changes based on the Minsk-2 with three modifications differing in structure of the additional equipment of input and output of the information. The Minsk-26 computer (N.A. Maltsev) and the Minsk-27 (V.E. Klotchkov) were to process telemetry coming from meteorological rockets and of the earth sputniks "Meteor". The Minsk-22 computers (V.K. Nadenenko) to which the Minsk-2 was connected, contained units of input and output of punched cards and also of the alphanumeric printing station contributed to a line of general-purpose computers.

With the Minsk-2/22 computer, it delivered an extensive standard program library (260 programs, 38000 commands), an autocode "Engineer" (AKI) (8000 commands), a system of symbolical coding (SSC), translators from languages of FORTRAN and ALGOL, the translator from language ALGEK (the language uniting properties of languages COBOL and ALGOL-60). A later delivery package included the translator from language COBOL and a data processing system "SAOD" (55 thousand commands). It was the richest package of programs delivered by the manufacturer with the computer in the USSR. In total, they produced 734 Minsk-22 computers, or 852 Minsk-2/22 computers. As a result, the lead positions of the Minsk computers in the country common park of computers had essentially become stronger.

In the mid 1960s, the Western countries produced large product lines of simple and rather inexpensive computers such as the IBM 1401, IBM-1440, and the Gamma-30 for business calculations. These computers featured a decimal-binary number system, word variable length and developed resources for logical processing of the alphanumeric information; these machines essentially raised the efficiency of processing of industrial and commercial information. The developers of the Minsk computers could not pass by the possibility to expand the usage of its products in the industrial sphere.

In 1966, the SDB factory developed the Minsk-23 computer intended for processes of data processing solutions for economical tasks, tasks of statistics, production management, and tasks of datalogical characteristics. The chief designer was V.V. Przhijalkovsky; the deputy of the chief designer was G.D. Smirnov.

The speed of the Minsk-23 achieved about 7000 operations per second. RAM capacity was forty thousand, eight-bit characters (bytes). A cycle of operation of the RAM and the computer was 13 microseconds. The capacity of the address storage (intended for storage of addresses of commands and operands, program and informational bases, and current addresses of an exchange with input\output devices) was 127 cells on nineteen bits. The number system was decimal-binary, a comma was fixed after low-order digit; the form of representation of numbers and commands was a character sequence of variable length. The quantity of addresses in the command was variable. Addresses of the commands utilized base and index methods. There were commands representing the whole procedures of data processing.

The Minsk-23 had a structure and the command system completely different from existing architectures at that time in the USSR and computers were oriented to do calculations of a scientific and technical nature. The Minsk-23 was the first domestic computer with character logic with a word and command of variable length. The computer had developed a system of interruption and suspensions and a universal link for peripherals. Actually, it used a byte-multiplex channel and a protected area of memory for service programs. One had the possibility of using a considerable quantity of index fields for each program array, special editing commands, and processing of fields of the variable length consisting of alphanumeric characters.

The Minsk-23 computer utilized a multi-program operating system that provided the possibility to run simultaneously three working and five support programs. Thus, it could work in eight directions with 64 peripherals simultaneously. A delivery package contained input equipment from punched cards (600 cards/s), input equipment from a punched tape (1000 strings/s), an alphanumeric printing device (400 strings/minute), a punch of maps (100 cards/min), and a tape punch (80 characters/s).

For the first time in domestic practice, the Minsk-23 used tape storage of rolled type with a bit density of 32 impressions/mm, which was completely compatible with the Western tape storage specifications. M.F. Chalajdjuk and A.M. Titov developed the storage system at the design office of the Industrial Automation with a brigade led by V.G. Makurochkin that made the project conform to industrial norms. For specialized systems using the Minsk-23 computer, they developed a device called "Blank", which read out formalized forms with pencil marks. The chief designer was V.K. Nadenenko; V.E. Klochkov and E.I. Mukhin developed the data communications equipment through telephone channels on a Minsk-1500 machine.

The developed a language of symbolic coding (SSC) from which they made the system software for the Minsk-23 computer. The library of system standard programs, in addition to the units intended for calculation of elementary functions, contained the programs, made carry out calls to input/output devices with code conversion and editing, to the sort utility, to the service program for a tape drive, and to the support programs. The compiler entered into a delivery package from the autocode, the machine-oriented language with a considerable quantity of the macros servicing input-output and a routine library. Additionally, the compiler structure included the loader to prepare loading units.

For the first time in the USSR, they designed the software structure for a small computer. The Minsk-23 included an operating system, which contained a batch processing screen monitor, a collector-loader, a link for the computer operator, a system for defining failures, and a coordinator for multi-program processing. M.S. Margolin headed the software development for the computer.

Based on the Minsk-23 computer, they developed an automated control system for some firms, including one of the largest in the country – the Novocherkassk electric locomotive factory. Unfortunately, the computer had no expected commercial success. They had made only twenty-eight computers that did not correspond to possibilities of a factory of G.K. Ordzhonikidze. Probably, it is necessary to consider that the main reason for commercial failure of this computer was that the user was not accustomed to it for its main concepts. The absence of compatibility with previous models, insufficient productivity on tasks of scientific and technical nature, and the poor development for firms and organizations that required processing of business information were some reasons for its failure.

5   The Minsk-3 Series of Computers

The Minsk-32 computer completed a series of Minsk computers. They developed the computer under the direction of its chief designer, V.V. Przhijalkovsky. In 1968, its production started.

The purpose of the new development was to create a computer that united the best features of the Minsk-23 and Minsk-22M computers with complete compatibility with the latest technology for data mediums and applications. The necessity of support of compatibility with the widespread computer Minsk-22M certainly constrained the possibilities of developing a new logical structure for the Minsk-32. However, the experience with the previous model had shown the importance of such compatibility for the user.

Compatibility support for the Minsk-32, the bit grid of the computer Minsk-22M consisted of 37 saved bits. It saved the formats of numbers with fixed and floating-point formats and with formats for all arithmetic and logical commands. Execution of these commands occurred in the same way as with the Minsk-22M. Programs fulfilled the input-output operations, interrupt systems, failure services, and responses to the operator. Thus, the Minsk-32 computer made hardware-software emulation of Minsk-22M programs.

The saving of the 37-bit grid complicated seriously the development of logical structure and functionality of the new computer. Nevertheless, it was possible to bring a variety of innovations raising overall performance and expanding functionality in logical structure of the Minsk-32 machine.

First, as an information unit in addition to the 37-bit words appeared a seven-bit character that allowed presentation of 128 code combinations, including the Latin and Russian alphabet. In a word, five characters took places, each addressed separately. Additionally, one could enter operating commands with sequences of bytes of any length; this was similar to the Minsk-23 computer. There were also commands for decimal arithmetic, matching, and editing. Secondly, the computer could receive commands in multi-program operating mode, which allowed one to handle four working programs at the same time.

Soon A.I. Bahir and J.G. Bostandzhjan developed a new economical ferrite storage with capacity of 65,536 bytes, with 38-digit words and with a cycle of five microseconds. They also developed logic circuits with new complex units of diode-transformer types and with a clock rate of 600 kHz. As a result, the average speed of the computer was 30-35 thousand operations per second, which is almost six times more than the capacity of the Minsk-22M machine.

The essential improvement was introduction in structure of the computer selector, the byte multiplex channels, and a system of universal links with peripherals (SUS VNU); this allowed the possibility of connecting to 136 different peripherals in the standard way.

The design used protection frames of memory area for each working program, an address storage device, an effective system of interrupts (suspensions, for the Minsk-23), and an electronic time sensor as part of the computer structure. A special commutator allowed uniting complexities to eight Minsk-32 computers for operations over a shared task.

The computer received the "dispatcher" program, which was an advanced tape-type operating system. With the computer, it was able to deliver a system of symbolical coding, a macro generator, and compilers for COBOL, ALGAMS, and FORTRAN languages. The total amount of delivered programs exceeded 500 thousand commands with eight thousand pages of the documentation. M.E. Nemenman headed the software operations; V.J. Pyhtin was the chief designer of the processor.

They manufactured the Minsk-32 computer up until 1975. In all, they manufactured 2,889 computers; as a result, the Minsk-32 became the most widespread general-purpose computer in the USSR.

Moreover, if one is to understand the meaning of a "general-purpose computer", it would be one that has equal or at least close productivity for scientific and datalogical tasks. We must admit that the Minsk-32, before the appearance of ES computers, was a unique computer of a general-purpose capability in the country. For its logical structure, this computer lost nothing to foreign computers of a similar class. In this case, the collective of the Minsk SDB at the end of sixty years was one of the most qualified in the country and the most prepared for the development of the ES computers.

6 Conclusion

The mass serial production of the Minsk-32 computer was possible thanks to intense creative operation of industrial and technological services of a factory of origin headed by I.K. Rostovtsev, S.A. Murygin, M.F. Chalajdjuk, J.V. Karpilovich, and A.M. Titov. The mechanization and the automation of production were possible due to a wide application of conveyer and product lines. Their constant upgrade and application of advanced technologies had made the Minsk factory of G.K. Ordzhonikidze, and then the Minsk production association, a leader of the domestic computer industry.

In 1970, the USSR proclaimed a state award to the collective of developers and manufacturers of Minsk computers for producing more than 4,000 computers, which was more than 70% of all computers in the country. The laureates were V.V. Przhijalkovsky, G.P. Lopato, J.G. Bostandzhjan, G.D. Smirnov, N.A. Maltsev, G.K. Stoljarov, I.K. Rostovtsev, M.E. Ekelchik, J.V. Karpilovich, and L.I. Shunjakov.

References

1. Korolev L.N.: Structures of the computer and their mathematical providing. PH. Nauka (1974)
2. Golubintsev, V.O., Kupaev, V.M., Sinelnikov, E.M.: Evolution of universal computers. The Soviet Radio, Moscow (1980)
3. Przhijalkovsky V.V.: A construction and service characteristics of computer Minsk-2. PH. Statistics, Moscow (1964)
4. Margolin, M.S., Nadenenko, V.K., Smirnov, G.D.: Electronic computer Minsk-22. PH. The Higher school, Minsk (1967)
5. Margolin, M.S., Skoromnik, M.G., Stoljarov, G.K.: Chuprigina L.G. Principle of operation of computer Minsk-23. Statistics, Moscow (1970)
6. Przhijalkovsky, V.V., Smirnov, G.D., Pyhtin, V.J.: Electronic computer Minsk-32 PH. Statistics, Moscow (1972)
7. Lopato, G.P.: Computer techniques in Belarus. IT and VC (January 1997)
8. Minsk, NIIEVM, MPO VT,
   http://www.Computer-museum.ru/Thecomputerfamily

Some Aspects on Computing
Means Development Philosophy

Jaroslav Khetagourov

Russian Academy of Sciences, Moscow, Russia
tshe@bk.ru

Abstract. Some factors influencing computer facilities and programming development appear in this paper. The negative role of present tendencies and development influence is noted.

Keywords: Computer facilities, software development, high-level languages, evaluation reliability, computer viruses.

1 Introduction

Computing means (CM) as the unity of computing devices (machines) and programming have been developing in the world for more than a half-century. During this period, scientists achieved fantastic successes in creating computing devices (machines) and in the development of programming.

Computing means have received broadest application in various fields of our life. CM development experience provides us with many opportunities for analyzing accepted decisions and considering various construction rationality estimations. The last is mainly defined by developing conditions and the aims of CM application.

Let's consider the connection between CM application aims and their development conditions. At the initial stage, CM application sphere was restricted by scientific and (mainly) military aims. That is why there were rather simple estimations concerning the rationality of decision methods. In this situation, we take into account two factors: the time of decision-making and the expenses. Therefore, using of CM in various fields of modern society has been essentially expanded and it has become more complicated. In addition, we have increased requirements to CM and estimations concerning its application rationality.

2 Computing Means and Storage Capacity

The historical development of computing means has shown that it was much simpler to realize the increasing hardware production task than to supply its application areas with programs for problems solving. It has strongly influenced CM development. Developers have begun to improve the basic technical characteristics of computing means (such as speed, memory storage) and productivity of programming (high-level languages, HLL).

J. Impagliazzo and E. Proydakov (Eds.): SoRuCom 2006, IFIP AICT 357, pp. 98–102, 2011.
© IFIP International Federation for Information Processing 2011

The increase of speed and memory storage capacity of computing devices was mainly provided by elements size reduction, which resulted in increasing its work frequency and reducing the sizes of devices in whole. It was also provided by increasing the number of simultaneously working circuits and computing devices.

The technological methods of increasing the fast-action processors apparently have reached a level of economic rationality. We estimate economic efficiency as the ratio between expenses for technological rigging and the output of a suitable design. On one hand, we can evaluate the market value and expenses for ensuring heat pipe-bend from big integration circuits (BICs) in connection with increasing of work frequency. On the other hand, we can compare the rigging of increased requirements concerning stability of BICs operating conditions for the reliability of the rigging. The increasing number of simultaneously working processors in connection with developing a crystal upon which some processors are placed, has received an additional benefit due to reduction of amount of external communications within the system.

3 Computing Means and Programming

The increasing of programming productivity provided by the large-scale application of high-level languages (HLL) takes into consideration the tasks and decision features for various application areas of computing means. The use of high-level languages essentially increases the output of programming tasks. However, the transformation of HLL program into the program for a computing device (i.e. program on the computer language, CL, program) led to substantial growth of its storage volume.

In summarizing the above information, it is possible to characterize the development of computing means as extensive. Accordingly, one could estimate its role in economic development of society. The extensive development of computing means has brightly revealed a positive feedback of its two principal parts: computing devices and programming. Therefore, the increase of program volumes targets an improved performance of computing devices; their perfection leads to the possibility of increasing the software. Thus, these two parts create conditions for each other's development that provides the existing parity in the distribution and development of works. At the same time, it generates economic interests of the sides.

As a result, the concepts of CM development keep the parity of income distribution between manufacturers of hardware and software. The extensive development of computing means negatively influences generation of new ideas and its introduction in CM construction.

4 Introspection

The important problem of computing means development is an estimation of a relationship between types and numbers of operations in HLL and CL and the choice of building computing means. Apparently, it is expedient to carry out analysis of movement from CL to HLL, assuming hardware realization of each step of this movement. The recognized advantages of HLL programming on the one hand, and the achievements in technology of element size minimization on a crystal on the other hand, plus the development of CAD tools that give conditions for the practical realization of complicated circuits of operation with HLL.

However, according to publications, this CM direction has not received essential development, apparently because of corporate reasons. On one hand, the marketing of software on CL was sharply limited, and on the other hand, the requirements of high-speed computing devices essentially changed (reduction of the number of CL commands more than an order, with the practical exception of operating systems).

It is necessary to note that application of CM with HLL hardware realization considerably reduces expenses of firms that create systems with computing devices for programming and debugging of programs of an entire system for its operation. However, these are problems for other groups or companies. Another important problem of CM development is the estimation of universality of computing devices for effective calculations of large-scale tasks.

For the overwhelming number of CM users (more than 80%), especially for CM working in various fields, the main thing is the minimization of operating cost. We can achieve this in two ways:

1. Application of universal computing devices and typical programs;
2. Application of specialized computing devices that effectively solve certain groups of tasks and apply the most convenient HLL.

The modern level of technology of computing devices, CAD application, creation of basic technical solutions (basing crystals for microprocessors and controllers) changes the assessment of the feasibility toward specialized decisions.

Leading CM manufactures are interested in production of various universal CM that have large series. It keeps extensive development and provides certain profit level and parity between the CM parts. Actually, for CM users it leans toward an increase of expenses. However, the interaction of CM producers is organized in such a way. It is one of the important factors constraining realization of new ideas.

5 Multi-processing

The use of crystals with several processors has created favorable conditions to increase the rate of calculations by increasing its number within systems in large-scale quantity. The decision problem to increase the output of many processors system has passed to algorithmers and programmers.

We can estimate the actual speed of system computing means by the ratio between the average of high-speed processor operations and the nominal sum of its action. We estimate the quantity of this ratio by a small number of peculiar properties that we determine by a kind of task and by the skill and intellect of the algorithmers and programmers.

The leading manufacturers of CM hardware and software support this direction; that is, it keeps parity of the income growth for these two groups. To reduce the influence of human factors, it is especially important to consider the direction of the automatic mis-paralleling of programs with the application of associative memory and current structures. (See the work of V. Bourtsev.)

Despite certain achievements, the problem of CM development regarding the task of ensuring the trustworthiness and reliability of output information continues to remain. As is known, we define trustworthiness of the information (absence of mistakes in output data) by the quality of used control methods of correctness of operating computing devices and the execution of the program by its testing.

It is necessary to note that we define the importance of requirements to trustworthiness of the information by the limited number of CM consumers (15-20%). However, the damage caused because of erroneous information could reach a quantity of nearly billion units. The way to achieve the requirements regarding CM trustworthiness and reliability is control. One way applies to hardware control methods (use of odd and even, module 3, codes, a majority method); the other way is applied to programming methods. Currently, hardware control methods extend the majority method by increasing the volume of the equipment and the consumption of CM energy by a factor of 3.5 to 4. Due to the use of universal CM, this decision is economically favorable to its producers. The application of programming methods based on the application of universal CM equipment, causes an increase of its performance by a factor of 2.5 to 3. This appears favorable for the manufacturer of the equipment and for the developer of software.

6 External Influences

Well-known methods of control state that neither the volume of equipment nor the increase of speed affects an increase in demand. These data again confirm that the existing economic relations between leading CM manufacturers constrain the creating and introducing new ideas of CM development. From a consumer's point of view, the cause of some incorrect information is the impact of viruses and various defects that cause damage to nearly a billion units.

Let's consider some features of the occurrence of viruses that break the order of program performance of any task. It is necessary to note that the occurrence of viruses within a program has not been connected with the infringement regarding the equipment operations of the computing device. The accepted measures concerning the elimination of a virus influence occur after its detection and they operate only on the virus. To predict the occurrence of a virus just now, according to the published data, is not possible.

In analyzing virus programs, we should note that within them, the commands transform the addresses and change the data in use. A possible number of similar combinations of commands within programs can cause practical harm beyond all bounds. Therefore, existing CM operating procedures are periodically broken by occurrence of the next virus whose influence is eliminated by the creation of the next anti-virus program associated with additional expenses for the user, plus the increasing of the program volume, and demands of speed requirements.

7 Conclusion

For elimination of viruses influence, it is necessary to enable the changes of the organization of a computing device and to stipulate the condition of the formation of a program. The realization of this direction demands essential processing, organization of work of computing devices, and programming methods that are associated with large expenses.

The economic relations between leading world companies have mainly determined the existing extensive order of CM development. Such relations constrain the application of basic and new ideas of development of computing systems. This situation can only change because of the appearance of new conditions and social needs.

The Algorithmic "Computer"

Zoya Alekseeva

Senior Research Assistant
Moscow Engineering-Physical Institute (MEPI)
zd_alexeeva@inbox.ru

Abstract. This article describes the software creation and development of the computer. Fortran 4 was chosen for scientific and technical problems. They named the language based on Fortran 4 language RTL (Fortran- real time language) treated the allocation of programs in the memory, different kinds of record commands in registers, and instruction structures in memory. It treated the principles of control units and developed an acquisition of some system statistics. Programming in the Fortran-like language increased the coding productivity fivefold in comparison to coding with an ordinary computer language.

Keywords: Algorithmic Computer, Fortran 4, high-level languages, real-time systems, real time languages (RTL).

1 Introduction

Development of computing means goes on two basic directions: development of the equipment and development of methods and means of the software creation. In the 1960s through the 1980s, software producing began at a high-level language (HLL) level. Thus for the creation of links connecting the program written in a HLL and a computer language, it was necessary to create a line of programs – compilers, which made the translation process. Usage of a HLL has certainly simplified the work of a programmer. However, the necessity of application of compilers has essentially increased the size of programs and consequently, has increased the time of performance of a problem; the physical size of the equipment of a computer has increased the time of entering of corrections at debugging.

At any moment of time at designing the specialized systems, the cost for the creation of the software began to exceed the cost of the creation of the equipment. Except for it, one other point appeared; if it is necessary quickly to change the program on the place of usage one mast have the technological equipment there it is not always possible.

The growth of the size of programs using HLL, translation and compilation for reception of a machine code in two to five times demanded an increase in storage size; it was also essential to increase the productivity of the "Computer" at the decision of problems in real time systems. All these circumstances have forced to search for new ways for the rational decision in total of the created problems.

Many language tasks have appeared in computer science because of the different kinds of problems the features of language required.

J. Impagliazzo and E. Proydakov (Eds.): SoRuCom 2006, IFIP AICT 357, pp. 103–116, 2011.

Therefore, they applied the Cobol language to the decision of problems of economic aspect, LISP for programs of processing of lists, Fortran for scientific and technical problems. The universal languages of priority were PL-1 and the Ada; however, within them appeared no means to address problems of the logical-managing side.

One of search variants of the structure of the "Computer" was a HLL computer language. In the 1960s, high-level source languages witnessed American publications about the "Computer" with hardware interpretation; in particular, Fortran-machines began to appear. The source language for the modified language was Fortran. The Fortran instructions executed in the machine and the device of management had nodes for processing each kind of instructions that transformed them to sequence of microinstructions. This development had not received wide distribution in view of great storage capacity of the equipment.

The idea of creating a "Computer" that worked on HLL has arisen in MEPhI in the beginning of 1980s at the chair of professor Ja. Khetagourov. Having made the analysis widely enough used at that time, HLL and a class of problems, characteristic for specialized real-time systems, came to a conclusion. It is rational to create a language similar to Fortran since for the algorithmic "Computer" it was simple to study, it was logically harmonious, and it was convenient for a wide class of problems and tasks. Having executed preliminary studies, materials had transferred in NPO "Agate" where, making a start from preliminary studies, it created the "Computer" that worked on a Fortran-similar language. Language expanded to include Fortran IV with inclusion of some operators of the PL/1 language. It included special operators and the descriptions allowing it to work real-time systems and with necessary conditions to protect programs and to process of the decision from malfunctioning.

For the aforementioned purposes, the operators used allowed the following:

o   Organize protection of programs and files,
o   Execute parallel work of programs,
o   Carry out blocking of interruptions,
o   Set reactions to interruptions.
o   Allow language means to fix situations interesting developers,
o   Simulate occurrence of signals of interruptions,
o   Distribute memory dynamical, statically or under the instructions.

To protect the workings of the machine against consequences of malfunctions, it helps to segment programs into different sites after entering the language; this allows the ability to make repetitions of the decisions in a case malfunctions situations. For example, one can monitor a condition of the "Computer" and the program storage when sensing the occurrence of a malfunction interruption. In addition, the common interrupt error library can organize with protection:

o   Task of priorities,
o   Introduction of several update wait operators,
o   Work with timer, etc.

To use this language, it has created the "Computer", directly realizing problems and the system software written on HLL language.

2 The Language Project

They developed the language based on the language Fortran IV and they named it RTF (Fortran real time). The set of RTF operators appear in Appendix A. Table 1 shows examples of situations of interruptions and reactions to them in a concrete system. (In some systems, other reasons can exist of interruptions and reactions to them.)

Table 1. RTF Examples

The name of a situation	Value
OVERFLOW	Overflow
UNDERFKOW	Loss of the importance
OVERTIME (E)	Time is exceeded (time interval specified in expression E or in a variable has expired)
KEY	Discrepancy of a key and the lock
SUBSCRIPTANCE	Output(Exit) of indexes abroad a file
CHECK (x1, x2, ...)	Reference (manipulation) to specified identifiers (e.g. programs, subroutines, labels)

Note that at the occurrence of the interruption signals, if a mask does not forbid it, the system reacts in the standard image if special reaction to interruption is not specified in operator ON.

3 Features of the Algorithmic "Computer"

Introduction of the high-level source language is natural and it demands one of two ways in the creation of software:

1) To process the program, translating it with a computer language (the compiler), and coordinating separate pieces in the concrete program in a single whole
2) To have as a computer language a high-level command language; that is, enter into the management of performance of operators the additional equipment for the analysis and performance of language designs that will distinguish such things as kinds of operands, operations, types of dates, and the order of their performance.

As mentioned above, the experience of developing the Fortran-like machine, where it carried out each operation on the processor, appeared impractical because equipment had a large physical volume. In the 1980s, integrated circuits appeared and they reduced the physical size of the equipment, and the idea has again caused interest since hardware performance of operators increased speed. We already mentioned the program advantages. However, requirements of the "Computer" have essentially grown on a range of numbers and their type such as on its memory size, which at that time was of low speed. In the developed "Computer", they have incorporated two important features: The numbers stored in the memory had a word length of 32 bits, and the memory size consisted of 65K words.

In the "Computer", words could consist of the following types:

o   Short number with the fixed point (16 capacity);
o   Number with the fixed point (32 capacity);
o   Number with the fixed point of double length (64 capacity);
o   Short bit line (16 capacity);
o   A bit line (32 capacity);
o   Number with a floating point (32 capacity);

Other features included the following. All numbers with the floating- and fixed-point representation appear in a complementary code. They used a bit representation for logic variables and applied logical values TRUE=0 and FALSE=1 in logic management. An attempt to execute operation with operands of different types is perceived as a mistake. If the operands involved in operation are different in type, the system will transform them as the more complex type; that is, the result will be with a floating point at addition of number with the fixed point with number with a floating point.

Expressions get in the arithmetic device in the arithmetic-logic kind, consisting of operations, variable with the indexes, set by the direct or indirect address, with the Polish record of arithmetic and logic operations, i.e. built on a priority. A basis of internal language is one of elementary records consisting of eight kinds. Furthermore, it stipulated four base registers for transformation of expressions. The address of memory is formed by the addition of number of pages (contained in the base register) and number of a half-word, which contains in an address part of kinds 1, 2, 7 records.

4   Kinds of Records

There are seven kinds of records as we now show.

(0) Record of the Zero Kind

000	Number of page
3p	13p

Establish the page number in the second base register

(1) Record of the First Kind

001	NBR	TYPE	Number of a half-word
3p	2p	3p	8p

Simple variable or constant addressing
(NBR = number of the base register, TYPE = type of a variable)

(2) Record of the Second Kind

010	NBR	TYPE	Number of a half-word
3p	2p	3p	8p

Variable addressed by indexing

(3) Record of the Third Kind

011	CI (code instruction)
3p	5p

Operation

(4) Record of the Fourth Kind

100	CI	Code of shift
3p	5p	8p

Shift operation

(5) Record of the Fifth Kind

101	CF
3p	5p

Designates codes of standard functions and managing symbols

(6) Record of the Sixth Kind

110	NBR	TYPE	OPERAND
3p	2p	3p	16-32

Direct operand

Table 2 shows an example of an assignment statement record in the memory of the system for the following Fortran-similar language.

$$A[I] = (B + C[J, R] * M[J, K]) * SIN(X) - 1.6$$

$$= 7.81516232431$$

Table 2. Records in memory

A	[	I	]
=	(	B	+
C	[	J	,
R	]	*	M
[	J	,	K
]	)	*	SIN
(	X	)	−
1.6			

Note that from the resulting record in memory we can see that the record is dense; we can superimpose blank bytes in cases of transition only at the beginning of another problem where the following operator is a label on which there can be a jump command.

5 Operations List Carried Out by the Arithmetic Processor

The arithmetic operations that execute on the algorithmic machine correspond to a set of operations of the usual machine; however, they are added with some operations of translation of one type of data to another. Table 3 shows this.

Table 3. Arithmetic Operations of the Algorithmic Machine

N	Designation	Name	Dates types
1	-	Monadic minus	1, 2, 6
2		Shift	1, 2, 3, 6
3	*	Multiplication	1, 2, 6
4	/	Division	1, 2, 6
5	+	Addition	1, 2, 3, 6
6	-	Subtraction	1, 2, 3, 6
7	ABS	Absolute size	1, 2, 6
8	SIGN	=1, if x>0; 0, if x=0;-1, if x <0	1, 2, 6
9	MOD	X=A+By, B-the whole, 0 <A <y	1, 2
10	.EQ	Equally	1, 2, 3, 6
11	.GT	It is more	1, 2, 3, 6
12	.LT	It is less	1, 2, 3, 6
13	.NOT	Logic NOT	4, 5
14	.AND	Conjunct	4, 5
15	.OR.	Disjunction	4, 5
16	.XOR.	Logic excluding OR	4, 5
17		Transformation 1 and 2 types in 3	
18		Transformation 1 and 2 types in 6	
19		Transformation 3 types in 6	
20		Transformation 6 in 1 and 2 types	
21		Transformation 6 types in 3	
22		Transformation 3 types in 1 and 2	

Table 4 shows a list of standard functions and managers of the symbols which are carried out in the arithmetic processor.

6 Instruction Structure

An instruction represents the operator, received after a translation of the command, which appeared in the high-level language. The length of the instruction can be any size, a multiple of half-bytes. Each following instruction begins with the half-byte following the end of the previous instruction. We make exceptions with the first program instruction and the instruction to which programs at performance management is transferred. Their record always begins with a new cell, irrespective of filling previous cells.

Table 4. Symbol Functions and Managers

N	Function	Name	Special cases
1	LOG (x)	ln x	Trashing
2	EXP (x)	Ex	Trashing
3	SIN (x)		
4	COS (x)		
5	TAN (x)	tg x	Trashing
6	ATG (x)	arctg x	
7	SQRT (x)		x <0
8	ATN (x, y)	arctg (x/y)	y = 0
9	x/y	Xy	x = 0 y <0
10	(		
11	)		
12	,		
13	{		
14	}		
15	=		
16	;		

The first instruction has the following structure:

1	1	Code of the operator	Quantity-in symbols			1st symbol				Sym bols			
				Attri bute			Following Attributes				Attributes		
				1 atr.			Following 2 atr.				A trace. 2 atr,		

The first byte of the instruction contains in two senior capacities an attribute at the beginning of the instruction, and in the others - a operation code. The second byte of the instruction includes a code of symbols number in the instruction and the attribute first symbol. Values of the symbols are the following:

    00 – Identifier of RAM
    01 – Identifier of ROM
    10 – Operator or a divider
    11 – Constant

The constant as it has been told above, can have length from one up to four bytes. To allocate under a constant standard number of bytes (4) the length of the instruction becomes in a significant amount of cases inefficiently large, with a significant amount on empty bites. To enter attributes of length of a constant, the analysis of the instruction becomes complicated. We have approached this in a different way on conditions of that time. Thus if in first attribute of a symbol was the attribute of a constant instead of second attribute, the attribute of a constant length was stated followed by the symbol of a constant, instead of following the attribute of the second symbol. If the attribute of a constant appeared in a place of second attribute, the code such as a constant appeared in the second half-byte of attribute in a place of first attribute. In this case, only one further symbol was read. For the symbols, which are distinct from a constant, the order of following became customary. The values of the attributes of a constant length are the following:

00 – 1 byte
01 – 2 bytes
10 – 3 bytes
11 – 4 bytes

The equipment determined the end of the instruction on zeroing account number symbols and on allocating the ends of a code of the instruction.

7   Features of the Managing Device

The performance of the operators in the machine at such internal language demands additional physical size of the equipment. Therefore, in addition to the usual elements of the management devices (for example, start an arithmetic device) in the algorithmic machine appears a set of devices, managing triggers, and counters. All these additional devices represent devices for retrieval and the preliminary processing of instructions. To accelerate the process, two registers act as buffers for consecutive reading the next two instructions words.

Preliminary processing concerns itself with six device entities:

- o   Arrangement tracking correct passage of the instruction from ROM (calculation number symbols and comparison originally determined)
- o   Device for forming the address of an operand
- o   Device for determining type of a symbol
- o   Device for determining type of a constant
- o   Microcode, organizing interaction of all devices DM
- o   Circuit of concurrence (determines concurrence of a symbol to code END and gives out an attribute of the instructions end of the instruction)

8   Some System Statistics

The experimental model of the "Computer" was executed on the 133 series. The arithmetic processor (without the controls circuit) consisted of 412 N-slot, from them on the managements device 138 N-slot were necessary. The time of performance of operation of addition for the first and second type of the data was 0.3 microseconds. For records of the sixth kind, it witnessed speeds from 0.7 up to 9.6 microseconds. Multiplication operations were carried out for records of first and second kind with speeds from 4.9 to 9.7 microseconds and for sixth kind from 5.2 to 10.1microseconds.

Logic operations were carried out for in 0.3 microseconds for first, second, and third kind records, and from 0.6 to 6.8 microseconds for sixth kind records. Logic operations executed in 0.1 microseconds for fourth and fifth kind records for operations such as inversion.

9 Summary Remarks

The results of development, debugging, and testing of work of an experimental sample on the decision computing and solving system problems, we came to the following conclusions.

1. Programming in the Fortran-like language RTL-77 has increased coding productivity in comparison with coding on a computer language by five times.
2. The programs written in the RTL-77 language, in comparison with programs written in a computer language, were reduced in size on the average by 3.3 times.
3. The memory size occupied with the program written on RTL-77 is approximately equal to a memory size occupied with same program written on a computer language, and three to five times smaller than the program written on HLL and compiled code on a computer language.
4. The size occupied by the equipment was 120 without package performance.
5. Speed of the "Computer" was about two million operations per second.

The departmental commission accepting the "Computer" recommended replacing the element base used in the computer by integrated circuits to reduce needful power and volume. However, because of weak development of integrated circuits at that time and the sharp reduction in financing in 1990, work had halted. Currently, this idea is still interesting for developers of computer equipment. The publications appear about development software without programmers with appearance technology "system on a chip" logical depth of operators and machine commands increase. It is desirable to develop method for data in computer in text or vocal forms.

Appendix A

Number of the operator	Operators, the alphabet, the basic concepts (formal record)	Explanations, examples
1 Input (F)	READ (a, n), READ (id, n), READ (a), READ (id), READ (a, w, n),	Ç – devices number-decimal number without a sign, n - number of operator FORMAT, - the identifiers list and cyclic elements (c), c - x = m1, m2, m3, where m1 initial value, m 2 - final value, m3 - a step. If a step = 1 it is possible to omit ид - the identifier of a file id1 id2 – the files identifier, records number. n - operators of input without a format (in the standard form - most frequently meeting in problems) w – the lists name of names of the variables subject to input \ to a conclusion (see operator NAMELIST) is similar to the operator INPUT
2 Write (F)	WRITE (a, n), WRITE (id, n), WRITE (a), WRITE (id), WRITE (a, w, n)	
	ALLOCATE (id1, id2) SET ALLOCATE (id1)	To make loading a file with name id1 for area of memory with name id2 (Transfer of the information between levels of memory, variables value A=1 when loading is ended.

Number of the operator	Operators, the alphabet, the basic concepts (formal record)	Explanations, examples
4 Format	FORMAT (b)	b - formatted expression. The operator is used for the indication of a format of the entered and removed dates and the accompanying text. b (b1 b2 b3 mb4, b5,), where bi - specifies, m – the specifies repeater or formatted group (b1, b2, m (b3, b4, b5,) b6..), m - the whole decimal number without a sign. FORMAT (// A. ...) - // - means the passing of 2 lines at a printout. FORMAT (nx, a1, a2...) n blanks before printing the first value or a symbol. b_i specifies can have type and a kind: I - integer, R - real, G –group printing the text, L - a logic variable, F - floating, X – the passing of a position, B - binary. Examples: mIw, mRw. W1, wGzzz---zi where "z" repeat w time, H'zz---z. m-the repeater, w1 - number of digit after a comma, w - the number of symbols (width of a field) FORMAT (E 12,4) - is entered - 824.0123 E-2 means: - 824.0123*10-2
4	The description of variables INTEGER (or begin with letters at absence of the obvious description: I, J, K, L, M, N) REAL DOUBLE PRECISION LOGICAL BINARY	The whole - 640 \0\00\-35 Valid - 04.\-700.\ .015\-.27У2 \6. ô-4 Double accuracy - 0. \ 6D-10\-10.2D4 Logic - .TRUE.\ .FALSE. Binary - 101. IB\-I. 001B\ - I0IE I-10B it is used in a combination to descriptions REAL, INTEDGER, DOUBLE, PRECISION
5	The alphabet: A I B I... IZ letters 0I1I2 I... I9 figures .T. I .F. I =, TRUE. I .FALSE E a designation of logic operations; designations of arithmetic operations: + I - I * I ** I / I +,-, x:; signs on operations of the attitude(relation): .GT. I .GE. I .LT. I .LE. I .EQ. I .NE. or>,>, <, <, =, =; designations of logic operations: .OR. I .AND. I .NOT. I or: "or", "¿", "not"; dividers:, I. I = I (I); indexes of following: GO TO I IF I CONTINUE I PAUSE I RETURN I STOP I WAIT; Descriptors and operators (see corresponding items(points))	
6	The Identifier	Any line of letters and the figures, beginning with the letter, no more than 7 symbols: B I F I A*B, ALFAITIME (identifier " TIME " concerns to the reserved identifiers)
7	LINE	- Any sequence of symbols, the prisoner in apostrophes, or with previously wHzz---z = ' zzz---z ' ' zz---z ' B - a binary line (from "1" and "0")
8	FILE A (i, j, r...)	A - the identifier of a file, i, j, r-identifiers of variables - indexes or constants
9	The description of file DIMENTION A (k1, k2, ... kn)	k1 - the whole decimal number - the top border of a files index. Dimension of a file = to number of indexes n.

Number of the operator	Operators, the alphabet, the basic concepts (formal record)	Explanations, examples
10 (F)	The operator appropriate A1... An = E	E - expression (arithmetic, logic) Ai - identifiers of variables by which (or to which) E.Primer's value it is appropriated(given): A1 = TIME - in a cell A1 contents timers value (circuit realization) are remembered
11	The operator of appropriate LABEL ASSIGN n TO m	m - the variable accepts value of a label (the integer without a sign)
12 (F)	GO TO n	The operator of unconditional jump
13	GO TO (n1, n2, ... nk)	Jump under the instruction (see operator ASSIGN)
14	GO TO (n, n, ... n), m	Calculated jump. m should appear in the left part of the operator of giving m = j, where 1 <j <k
15 (F)	IF (E) n1, n2, n3	The conditional arithmetic operator of jump GO TO n1 - if E> 0 n2 - if E =0 n3 - if E <0
16	IF (L) Q	The conditional logic operator. L - logic expression, Q - the operator - not D0 and not IF
17 (F)	THE OPERATOR OF CYCLE DO n x = m1, m2, m3,	m 1, m2, m3 - initial value, final value and a step accordingly for XX. If m 3=1 it is possible to not write it(him). The cycle goes up to the operator with a label n inclusive. This operator cannot be IF, DO, RETURN, STOP.
18	CONTINUE	The operator of continuation. Carries out jump to the following operator. One of operators replaces at inadmissible connection of adjacent operators
19	PAUSE n	Stop the working program. By pressing button " Start-up " - work proceeds from a place stopping. The label n is given out on the consol
20	STOP n	Stop without an opportunity of start, only button " Start-up "
21 (Rt, pairs)	WAIT (E, k)	E - the variable of type the INTEGER - specifies real time, which is necessary to wait before to continue a course of the program (following for WAIT the operator). To - a constant such as INTEGER, specifying required accuracy of readout of an interval in мкsec
22 (Rt, pairs)	WAIT b1, b2, b ... bi	bi - the logic variables, following for WAIT, the operator will be executed, when all bi begin TRUE
23 (Rt, pairs)	WAIT (ид R, A)	IdR - the identifier reserve the register addressed. And - a variable of type binary (BINARY). The subsequent operator will be executed, when (idR) = A Is checked at each interruption
24 (Сис)	Assignment of priority PRIORITY (I=E)	I - the identifier of the program or a branch (TASK), E - a constant or an integer variable.
25	Descriptor of branch TASK and (A, z, v)	The description of a problem, the identifier, A - a variable of type a priority, z - a variable such as lock-the lock (can be absent). v - a variable such as MASK - a mask.
26	Descriptor of program PROGRAM (I, z, v)	a-the identifier of the program, comes to an end END. z, v that, as in 25, can be absent

Number of the operator	Operators, the alphabet, the basic concepts (formal record)	Explanations, examples
27	ATTACH (I, y, r)	A name of a branch (TASK) which can go in parallel with the causing program (if there is a corresponding equipment). I - value of a priority at the moment of input (the number should be appropriated). R- a variable such as EVENT - a condition of inclusion of a branch. Can be absent R and V. y - a variable such as KEY a key.
28 (F)	CALL a name, Y (n1, ..., nk)	Start of procedure, procedure of function of the subroutine with a name. n1, nk - values of actual parameters, y - the variable such as KEY, can be absent.
29 (Sis)	ON a b c	The operator of the reactions task interruption. a - a name of one of the situations accepted in system; b - the identifier (SNAP) which can be absent. If it is, the information on a condition of system is given out to the operator at the moment of interruption; C-the operator (for instans GO TO M), specifying reaction of system to interruption - the program with a name "æ" or the subroutine will be started. If "c" is absent, the system reacts in the standard way.
30 (Sys)	REVERT a	The cancellation of reaction to interruption (on ON a ...) also restores the reaction specified in the covering block: ON b ON a REVENTA.. Operates b
31 (Pairs)	KEEP a (r1, r2, ... rn)	The operator who is carrying out preservation of values of a condition of the program a for variables, whose identifiers are specified in (...), there can be names of registers and time
32	FREE a (Y)	To clear area of the RAM from the information which have been written down in area with a name "a." Y type KEY.
33	FREE (a1, a2, ... an)	To clear the RAM from the variables specified in brackets.
34	COMMON Ir1 I A, B, C Ir2 I R, T, Z ...	The operator allocates the common memory with a name r1 for variables A, B, C, (they will be stored(kept) in one area of memory by way of the list), with a name r2 for variables R, T, Z, etc. If the common area is one, it can be not marked. Identifiers in the list can be identifiers of files, but then their characteristics should be set in the operator of description DIMENTION, or in operator COMMON Ir1 I A, B (10, 5), C (see item 9)
35 (Pairs)	EQVIVALENCE (r1, r2)	The variables specified in brackets of operator EQVIVALENCE, will be placed on the same place (in cells) memory. For example: at DIMENTION (r1 (53), r2 (15)) files demand an identical place, but they should not be used simultaneously except for a mark of the same file by different identifiers in different programs for reading.

Number of the operator	Operators, the alphabet, the basic concepts (formal record)	Explanations, examples
36	FILE a1 (m1, l1, t1, z1,), a2 ...	The description of a file. a1, a2, ... identifiers of files or libraries (LIB), mi - quantity of records in a file a i; li - the maximal length of record in a file, ti - the maximal length of the record following on the order after the termination(ending) of reading or record; zi - a variable such as lock.
37	OPEN a1 (Y1), a2 (Y2)	To open files a1, a2, ... - on AM (auxiliary memory. Yi - variables such as key
38	CLOSE a1 (Y1), a2 (Y2)	To close files a1, a2, ... - on AM Zi Yi can be absent in 36, 37, 38
39	ALLOCATION (a1)	Gives out value '1' B, if the variable a1 is placed in the RAM (ALLOCATE (a1)) and '0' In - otherwise.
40	Descriptors of accommodation of the data in memory AUTOMATIC (a1, a2, ...) STATIC (a1, a2, ...) CONTROLLED (a1, a2, ...)	If there is no description for type of accommodation of a variable type AUTOMATIC is meant. Accommodation of variables in the RAM before performance of the program invariable before its(her) end. Accommodation dynamic occurs before each input(entrance) in a segment (in sense Fortran the common) or in the block (a LIFO principle) It is control operator ALLOCATE
41	NAMELIST |w1 |, A1, B1 ..., | w2 |, A2, B2, ..., |wi |, Ai, Bi, ...	wi - the name of the list of names Ai, Bi, ... wi is used in operators WRITE, READ for reduction of record of entered or deduced(removed) variables.
42	DATA a1, a2, (n1, n2, ...,)	a1, a2, ... - the names list of variables by which numerical values n1 are appropriated, n2, ... by way of conformity. The element ai can be an element such as a cycle: (ai (I)), where I=m1, m2, m3 .... For example, If ni repeats some times successively enters the name so: m*ni - number of repetitions on each m-cycle.
43	The subroutine - function FANCTION B (x1, x2, ...)	t - or it is empty, or a descriptor such as function, B - a name of function, (x1, x2, ...) - the list of formal parameters. The result will be worn out on identifier B. In (a1, a2, ...) can meet in operators, that causes calculation B for values of actual parameters (a1, ...) the body in words RETURN and END.
44	The Subroutine (procedure) SUBROUTINE P Y (x1, x2,..)	P - a name of the subroutine Terminates. The call goes on operator CALL P (n1, n2, ...). Comes to an end RETURN and END. Y - variable such as mask (in 43 and 44 can be absent).
45	Descriptor EXTERNAL a1, a2, ...	This descriptor is used in case arguments of function or the subroutine can be names of functions or subroutines. In this case in the basic program the descriptor (45) in which names of external functions and procedures which should be handed as actual parameters are listed is included.

Number of the operator	Operators, the alphabet, the basic concepts (formal record)	Explanations, examples
46	The operator simulating a situation of interruption SIGNAL a	á - a name of one of situations of interruption which can arise in system. The operator simulates occurrence of a situation "á" and transfers management to operator ON working in the given area of the program. If the situation specified in a, is not included, that is operator ON where there is a parameter and interruptions will not be is not executed yet. Management will be handed to the following operator for SIGNAL a. This will take place and after the termination of processing of interruption ON .If a case of performance of operator SIGNAL a situation CONDITION (a) and if operator ON a, b, c has instead of and operator CONDITION (a) follows as it mentioned above, interruption is developed.
47 (Def)	The description of "lock" LOCK Z (n)	z (n) - the file of binary numbers being "lock" which should be open by "key" KEY Y (n) (see 48). "Lock" is established in programs, files and branches.
48 (Def)	The description of "key" KEY Y (N)	Y (n) - a file of the binary numbers representing "key". It(he) is compared to "lock" LOCK z (n)
49	The operator of a supply of file REWIND a	The file with a name "a" in a condition when 1-n record can be read out or be deduced(removed) is established.
50	The operator of return BACK PACE a	Return from the current record of a file to a place of interruption is carried out.
51 (Sys)	The operator of search of record of file FIND (an)	Record of a file "a" with number n is established in position when input-output of this record will demand minimal time
52 (Sys)	To place in library PUT INTO LIB (r1, r2, ... rn) (m1, m2, ... mn)	To place variables from the program with the specified name in which there was operator PUT INTO LIB, r1, r2... rn, in library in cells m1, m2, ... mn,
53 (Sys)	To take from library GETEROM LIB name Y (m1, m2, ... mт) (r1, r2, ... rт)	To choose from library the program with the identifier "name" and contents of cells m, m, ... m to place in cells r, r, ... r
54 (Sys)	Reservation and the name of a place in RAM AREA a name, z (n, b)	At performance of this operator in the RAM the area from n words (n - decimal number) if the OS will find enough place not closed by operations system (or more high priority the program) and if it is not specified "b" - the identifier of a variable which accepts value of the physical address of a cell of the RAM is allocated(removed). If "b" is present at operator AREA OS checks availability of area in the size "n", since a cell "b".
55 (Sys)	MASK V	The description variable V as a binary variable for masking the register of interruptions.
56	CHECK (x1, x2, ...,)	Interruption of performance of the program at an output on identifiers or the labels specified in the list. Start-up from the consol.

Academician Andrei Ershov and His Archive

Irina Kraineva and Natalia Cheremnykh

A.P. Ershov Institute of Informatics Systems of the SB RAS
Novosibirsk 630090, Lavrentiev ave., 6
{Cora,cher}@iis.nsk.su

Abstract. Andrei Petrovich Ershov (1931-1988) was a mathematician and specialist in the field of programming theory and automation. A graduate of Moscow State University (1954), he worked in ITMCT AS (Academy of Sciences) USSR, in the AS USSR Computing Center. From 1961, he started working in the Siberian Branch of AS USSR (SB AS USSR): first in the Institute of Mathematics, from 1964 as the head of the Programming Division of the Computing Center, and then as head of the Experimental Informatics Laboratory. He was a doctor of Physical and Mathematical Sciences since 1967, a full member of the AS USSR since 1984, and professor. He was a member of the Association for Computing Machinery (ACM), a Distinguished Fellow of the British Computing Society, and vice-chairman of the IFIP Algol Working Group.

Keywords: history of informatics, academician Ershov, Siberian school of programming, optimizing translators, ALFA algorithmic language.

1 Introduction

The role of Andrei Ershov in the establishment and development of system programming in Russia cannot be overestimated. It is enough to say that it was under his guidance and immediate participation that the first optimizing translator from the Algol-like ALFA algorithmic language was developed; he is justly considered as the founder of mixed computations. Ershov was the first to start experimenting with teaching programming at schools; he authored the famous thesis "programming is the second literacy".

A.P. Ershov passed away very early. He was not even sixty, but his creative heritage still causes lively interest. One of the reasons for that is the unique archive containing over five hundred folders with documents reflecting the path of the academician and history of the development of informatics in the USSR. It sometimes seems that Andrei never in his life threw away a single piece of paper. Massive correspondence exists including over 5,000 letters, manuscripts of books and articles, reports, reviews, diaries, memos, even airplane and cinema tickets, invoices, even menus from ceremonial dinners and other material evidence of past events in different

J. Impagliazzo and E. Proydakov (Eds.): SoRuCom 2006, IFIP AICT 357, pp. 117–125, 2011.

parts of the world. Ershov carefully collected and systematized all this both chronologically and by subject.

Upon Ershov's untimely death, his archive, together with his remarkable library, was passed by his heirs to the SB RAS Institute of Informatics Systems. The value of this unique archive was as obvious as the necessity to process and systematize its documents and put them into broad scientific circulation. The conventional tools for archive processing are considerably limited, which gave birth to the idea of using modern methods of electronic document processing and their presentation in the internet.

2 Building the Archive

In 2000, with the financial support from Microsoft Research, the work on the creation of the electronic version of the archive began.[1] The work is performed by the employees of the Institute of Informatics Systems and the xTech company, and is supported by RFBR and RHF grants; UniPro and Atapy Software companies also contributed to the project. Currently, the work is close to completion. Historians of science are gaining access to the immense systematized bulk of documents that they can directly place into scientific circulation.

The documents in the archive were systematized by Ershov himself by subject-chronology principle and collected in folders devoted to certain subjects. They include materials from his trips to international and national programming forums, correspondence, documents related to research projects, drafts of articles and other scientific works, memos, scientific council meetings agendas, various official documents and much more. The structure of the electronic archive was prompted by this system; all documents were distributed by subjects that reflected consecutive steps in Ershov's life path; they showed different aspects of his activities. The archive is divided into twenty-eight subjects, which in turn contain groups and subgroups of documents united by common subject.

The earliest documents in the Ershov Archive date back to the times of his study in the 37th Boys School of Kemerovo. These are very curious biographic entries: several diplomas for excellent study and exemplary behavior, diplomas for winning sports tournaments, grades chart for the 1947/1948 academic year with only excellent marks, a draft of the written test in algebra at the exam for the school-leaving certificate. In the senior years at school Andrei took active part in extracurricular activities; he worked in the school Komsomol organization. In the autumn of 1947 the Komsomol regional committee awarded him with a tour to Moscow to the celebration of the capital's 800th anniversary and 30 years of the October Revolution. The Archive retains programs of shows in Lenkom (Lenin Komsomol Theatre), Mkhat (Moscow Art Academic Theatre), TsDT (Central Children's Theatre), Obraztsov Doll Theatre, tickets to Conservatory - evidence to the broad cultural program of the journey [1].

Several other folders contain notes and summaries of lectures from Ershov's student and postgraduate period. Among those is a modest notebook with eight lectures from the legendary course "Principles of Programming" delivered at the

---

[1] http://ershov.iis.nsk.su/ershov/russian

Mathematics and Mechanics Department of the Moscow State University by Alexei Andreevich Lyapunov [2]. The first lecture took place on October 29, 1952, in the middle of the semester. Ershov recalls that these lectures were not regular university lectures, which a professor reads using summaries written well beforehand. Those were an improvisation. "We later realized, - Ershov wrote - that at the beginning of the course Alexei Andreevich knew about programming not much more than we did at the time. In a sense, he learned together with us. But these bits of knowledge, multiplied by his brilliant intellect and immense general and mathematical culture, allowed Alexei Andreevich to seize the fundamental nature of programming and create its methodology "[3].

The archive's documents reflect the difficulties of the preparation and defense of Andrei Ershov's PhD thesis. In 1958, his famous monograph, Programming program for fast electronic computing machine, was published. The book was printed in 4000 copies and appeared in bookstores on the 13th of August; a month later, on September 12, Ershov wrote in his diary: "PP is out of stock in bookshops! A pleasure for the author" [4]. Perhaps Andrei could have presented the book as his PhD thesis; however, he chose to write a more "mathematical" work. A.A. Markov, a well-known mathematician whom Andrei Petrovich deeply respected, agreed to be Ershov's opponent. It is clear from the diary notes that the work on the thesis went slowly and with difficulty, because Markov detained the manuscript for long periods. Nevertheless, Andrei considered all his remarks with great attention:

> "Spent the day, excluding a short Sunday working-picnic for planting trees, making corrections to the dissertation according to Andrei Andreevich's remarks. He made twenty-eight remarks in the first fifty pages. Most of those can be easily taken into account. Despite that he let me down somewhat with timing, I am very grateful to him. After his reviewing the work will be almost flawless."

> "Continued proof-reading the dissertation. Wrote a proof to the theorem on the connection between operator algorithms and graph schemes. In the evening talked to A.A. Markov. He strongly disliked programming expressions about "changing" programs, variable commands, etc. He proposed to exclude the parts dealing with the comparison of computational and conventional algorithms from the introduction, making it much less "programming". He is right in some points, especially when he talks of the fuzziness of programmer terminology, but he seems to misunderstand some things" [5].

In the end, A.A. Markov resigned from his responsibilities; Andrei Ershov defended his dissertation already in Novosibirsk, in 1962. His opponent was a famous algebraist, academician A.I. Maltsev.

From 1961, Andrei Ershov lived and worked in Akademgorodok, a suburb of Novosibirsk, where he moved on the invitation of one of the founders of SB AS USSR, academician S. L. Sobolev, director of the Institute of Mathematics. Andrei was to become head of the Algorithms Theory Laboratory of the institute. The archive contains documents reflecting the selection of staff, the research progress, and the everyday life of the newborn scientific center.

3 The Alpha Compiler

The history of informatics in Siberia began, without doubt, with the beginning of the Alpha compiler project. Documents in the archive allow to trace its history. The first large project, which started under the supervision of Andrei Ershov first in the Institute of Mathematics and continued later in the Computing Center of the SB AS USSR, where Andrei transferred with his Programming Division in 1964, immediately grabbed the attention of world scientific community. In 1966, Ershov received a letter from K. Levitin, the editor of Znanie-Sila ("Knowledge is Power") magazine, where Levitin wrote: "Last Wednesday - on the 30th of March - M. A. Lavrentiev gave a talk in the Central House of Literature about Akademgorodok. While answering a question he mentioned your name - in the following context:

> "We have a scientist, Andrei Petrovich Ershov, - he said - whom the Americans have been trying to entice for a long time. They invite him over to work - for a year, or two, or for as long as he likes, together with the family, on the most attractive terms - with a 3000$-per-month salary and additional payments for consulting. In our country, he is a mere PhD" [6].

There are several thick "log books" in the archive, each reflecting a stage in the creation of the Alpha compiler. They are called accordingly, The *Birth of α-Compiler*, *The childhood of α-Compiler*, *Puberty of α-Compiler* [7]. Notes were added almost every day. At first, the logs reflected the project progress, and later, when debugging began, log notes began to play an even more important role. The developers briefly summed up the results of running a program, marked mistakes, passed information to colleagues who took shift after them in the computer room. Other final, preliminary and intermediate reports on scientific research projects (there are over 30 of them in the archive) reflect research, production and science managerial side of the activities of research groups headed by Andrei Ershov.

A. P. Ershov was one of the most "out-going" Soviet scientists – he made over fifty foreign trips in the course of thirty years of his active scientific career, although the number of invitations he received was much greater. However, Ershov's first trip to the USA in 1965 could have become his last one. He then gave a talk devoted to the Alpha-compiler at the IFIP Congress. Americans showed great interest in the state of affairs in the Soviet computer technology, and arranged Andrei an additional trip to San Francisco and Los Angeles. Based on Ershov's talk given at the Los Angeles ACM division, an article by R. Henkel appeared in the Electronic News, called "Soviet Expert on Soviet Machines: Not Enough and Not Too Good". The article caused violent feedback in the motherland: the President of the USSR Academy of Sciences, M.V. Keldysh, sent an angry letter to G.I. Marchuk, the director of the SB AS USSR Computing Center at the time, in which he accused Ershov of distortion of information about computer technology in the USSR and even of unauthorized disclosure. Unfortunately, we were unable to find that letter, neither in SB RAS nor in RAS archives. Ershov's reply, however, survived. While essentially answering the accusations, he simultaneously expressed many remarkable ideas concerning cooperation with the USA in the area of computer technology [9]. This document is a unique testimony of Andrei's ability to analyze clearly the situation, the strong character of the still very young man and his skill to stand upon his rightness. The scientific reports on Ershov's foreign trips, whose drafts remain in the archive, were repeatedly published both in magazines and as separate brochures, always being a source of new knowledge and fresh ideas [10].

4 Comments and Correspondence

Ershov attended a great number of international conferences, congresses and seminars, and the archive contains materials about their organization and holding: invitations, characteristics from party committees, which were an indispensable for anyone wishing to travel outside the country at those times. In this respect the story of the organization of the International Colloquium on Mixed Computations, which took place in Denmark in 1987, is a telling example. A Danish scientist, D. Bjorner, organized the colloquium. He managed to find considerable funds for the support of soviet scientists who achieved notable results in the area of mixed computations that caused broad interest among specialists worldwide.

Documents in the archive allow us to trace the formation of the soviet delegation and the difficulties caused by party officials that it had to cope with. There remain letters of Andrei Ershov to academician E. P. Velikhov, who at the time was head of the AS Informatics Division and whom Andrei addressed as the ultimate authority. Several members of the delegation were denied permission to leave the country; among them was S. S. Lavrov, head of the Institute of Theoretical Astronomy, who received a party penalty shortly before. Ershov wrote to Velikhov:

> "Evgeni, I strongly believe that the situation with the trip to Denmark is the case when one should separate events in the institute and Lavrov's directorial duties from the advisability of his participation in the seminar on mixed computation. We shouldn't hinder this worthy and important event. I am asking you to support us by calling comrade Fateev[2]" [11].

As a result, Lavrov was able to attend the colloquium, yet some of the candidates for the trip did not pass the "party control".

The archive contains remarkable examples of biographic genre: essays about Ershov's close colleagues - talented, but deceased untimely: Gennady Kozhukhin, with whom they began work on the input language for Alpha-compiler, Gennady Zvenigorodski, a wonderful programmer and teacher, who put his energy into teaching programming to schoolchildren. Andrei wrote many heartfelt warm words about his colleagues and teachers: A.A. Lyapunov, G.I. Marchuk, E.V. Dijkstra, and others. These works portrait the images of first programmers as well as show the directions of scientific research of the period. When Yu.I. Manin read the essay about Dijkstra, he remarked:

> "The text is full of natural and kind feeling. In the strange atmosphere of hostility and aggression, which is getting so dense in the professional communities of Moscow that I belong to, living becomes cramped and sultry. I recall remorsefully my own silent participation in the sittings of "Uspekhi MN" editorial staff, where "problems" such as lack of permission to publish anniversary articles in honor of A or B, or the question of whether we could call C outstanding when D had already been called famous, were discussed – and other rotten nonsense of the kind." [12].

---

[2] Presiding Commissioner for Foreign Trips of Leningrad regional committee of the Communist Party.

Andrei was truly interested in the history of science, and not only computer science: He recorded interviews with a prominent Russian economist, academician Aganbegyan, the Polish computer scientist V.M. Turski, a, the president of the SB AS USSR, academician M.A. Lavrentiev. Parts of the interview with Lavrentiev were used in Ershov's and M.R. Shura-Bura's joint work, *The Establishment of Programming in the USSR*. Curiously, the manuscript mimics Lavrentiev's manner of speech so authentically that it reminds a transcription of a tape recording:

> "We needed twenty-thousand lamps. But they would give only five thousand to the whole Academy. We tried this and that, but then got an idea. So we go straight to the radiotechnics people and say, how's your military acceptance, doesn't it bother you? You don't ask, they answer, we're wasting all our time on that, how do you check them? And we say, let's make a deal. You make us a revolving fund of 20 thousand lamp, we install them, record modes and give all data back to you. They were only happy to make a deal, so we got our lamps. Parshin[3] was mad: "A-holes! Those lousy academicians are cheating us!" [13]

Ershov's voluminous correspondence is especially interesting. The archive contains about five thousand letters Andrei sent or received. People addressed him with all sorts of questions. Different people wrote to Andrei, and he answered many of them personally. There are, for example, numerous letters from the period of introduction of informatics into the school course. This innovation had many proponents and just as many opponents. Among those who wrote were baffled schoolchildren, to whom the new subject seemed too difficult, and teacher worried by the lack of required training. There were, however, many letters that Ershov answered with pleasure. Those include a very serious letter from a high school student asking which college to choose in order to become a qualified computer science specialist [14]. Andrei maintained long-lasting scientific and friendly correspondence with many outstanding scientists such as D. Knuth, J. McCarthy, E. Dijkstra, Yu.I. Manin, and V.M. Glushkov as well as business correspondence with academicians M.V. Keldysh, A.A. Dorodnitsyn, A.I. Berg, and others.

5 Non-computing Areas

Since most of Andrei's life was connected with Akademgorodok, many documents reflect that special atmosphere which formed here thanks to the concentration of intelligent and multi-talented personalities. Life-science discussions on the subject "Life of wonderful ideas" in the SB AS House of Scientists attracted numerous participants and listeners who eagerly witnessed the progress of scientific research. There remain materials of Ershov's public speeches, when he took part in discussions with the audience. Notes from the audience in the archive speak of the sincere interest of the public to the new branch of science – the creation of artificial intelligence: "When will a working model of cognition be constructed?" "Do they use the achievements of physiology of higher nervous activities and psychology in the

---

[3] P.I. Parshin. The Minister of the USSR engineering and instrument-making industries (1946-1953, 1954-1956).

creation of artificial intelligence?" "Won't the creation of artificial intelligence pose any danger to mankind?" [15]

The archive contains several short essays Andrei wrote under the impression of casual observations. They testify to Ershov's undoubted literary talent, convey impressions and moods and truthfully represent the spirit of the moment. Once, walking on the shore of the Ob Lake, Andrei saw academician A. M. Budker, director of the Institute of Nuclear Physics, skating. Andrei wrote about this bright memory after over twenty years:

> "It was in the year 1961 or 1962. November. The Ob Lake froze, but the snow hadn't fallen yet. A natural skating rink formed. A group of people on the ice, dressed in winter clothes already. And there was Andrei Mikhailovich in a whitish scarf, skating alone. I watched him for five minutes or so. He definitely felt in the focus of events. One could see he was happy about this early winter, and was feeling the swift ease of skating once again, youthfully. I recall this scene again and again, and it conjures up an image of perky, brisk beginning of a new life." [16]

The image of the archive wouldn't be full without mentioning the documents which characterize the personality of Andrei. In July of 1987, Ershov was elected to the Organizing Committee of the V. I. Lenin Novosibirsk Sovietski District Children's Fund. On October 14, 1987, he took part in the work of the Establishing Conference of the Fund as an Executive Board member. Shortly before that, on the 12th of October, Andrei transferred 1000 rubles to the Fund from his personal savings. Andrei was deeply touched by the problems of orphaned children, and he repeatedly addressed the chairman of the Novosibirsk Sovietski District V. V. Generalov, asking him to support the initiative of Z. V. Borodayevskaya, a citizen of Akademgorodok, for the establishment of a family-style orphanage and provision of a cottage in Akademgorodok [17]. The decision was made in March of 1989, already after Ershov's death, when Zoya Vladimirovna and her six foster children moved to a house on Zolotodolinskaya Street. This family of ten still thankfully remembers the invaluable help from Andrei.

6 Conclusion

Ershov was a jovial man, always the soul of the party since the university years. He played guitar, sang, wrote poetry; he loved theatre, books and music. A good song never left him indifferent. Looking through the notes Andrei made in the course of his preparation for the exam on the theory of differential variable as a postgraduate, we found a handwritten text of a song from the film called *Vernye druzia* (True Friends), put down on paper hastily right after it was broadcast on the radio. It was April of 1955... [18]. During his trip to England in 1958 to the International symposium on mental processes automation he attended a concert of Royal Philharmonic – one of the four most important orchestras of Great Britain, and saved the program carefully [19]. He was lucky to listen, among other things, to the Haydn's *London Symphonies*, performed by an orchestra conducted by Sir Thomas Beecham – the piece specialists consider to be the most masterful achievement of this conductor. As a souvenir he brought sheetmusic and lyrics of "Happy birthday" written down by Patricia, the daughter of V. Bricks, employee of Elliot Brothers (London) Ltd [20].

One can find in the archive manuscripts of practically any of Ershov's articles and monographs. Two voluminous folders contain articles and materials reflecting different points of view upon the subject of informatics [21]. Perhaps Andrei collected these materials for another article ors book. A very interesting detail are the numerous notes with marks such as "Ideas and wishes in store"or "Idea!" [22]. Not only historians of science but also actively working researchers may find useful information in these manuscripts, and maybe even borrow some new ideas or find a creative stimulus in Andrei Ershov's the old but still relevant works.

Acknowledgment. The authors acknowledge those who have helped with the sources and development of this article. In particular, they are appreciative of the support provided by RFH, project № 05-03-12304.

References

1. A.P. Ershov's Archive. Memorabilia from the tourist trip to Moscow in (November 1947), http://ershov.iis.nsk.su/archive/eaimage.asp?did=13292&fileid=130921
2. Notes of A.A. Lyapunov's lectures Principles of programming, http://ershov.iis.nsk.su/archive/eaimage.asp?lang=1&did=31044&fileid=164717
3. Ershov, A.: In memory of Alexey Andreevich Lyapunov (talk at the memorial session of Siberian Mathemaical Society, pp. 243–245 (October 15, 1973), A.P. Ershov's Archive - Folder 532
4. Department Head's Journal, A.P. Ershov's Archive - Folder 35, p. 104
5. ibidem, p. 111
6. Letter from Levitin, K., to Ershov, A. (April 2, 1966), http://ershov.iis.nsk.su/archive/eaindex.asp?lang=1&did=20004
7. Journal ALPHA compiler: the birth. Events log (22.05.63–29.10.63), http://ershov.iis.nsk.su/archive/eaindex.asp?lang=1&did=2757
8. Henkel, R.: Soviet Expert on Soviet Units: Not Enough and Not Very Good - Electronic News (July 07, 1965), http://ershov.iis.nsk.su/archive/eaindex.asp?lang=1&did=26958
9. Letter from Ershov, A., Keldysh, M. (July 20, 1965), http://ershov.iis.nsk.su/archive/eaindex.asp?lang=1&did=20854
10. Ershov, A.P.: Computer Science in the USA. Based on the trip to the 3rd IFIP Congress, USA (May 25-29, 1965), Moscow, Computing Center of USSR AS, p. 339 (1966), http://ershov.iis.nsk.su/archive/eaindex.asp?lang=1&did=12425
11. Letter from Ershov, A. to Velikhov, E. (July 25, 1987), http://ershov.iis.nsk.su/archive/eaimage.asp?did=3046&fileid=80959
12. Letter from Manin, Y. to Ershov, A. (Januray 15, 1981), http://ershov.iis.nsk.su/archive/eaimage.asp?fileid=77610
13. Ershov, A.P.: Interview with academician M.A. Lavrentiev (October 26, 1967), http://ershov.iis.nsk.su/archive/eaimage.asp?did=17909&fileid=137068

14. Letter from Ershov, A. to Rechkalov, I. (March 10, 1986),
    `http://ershov.iis.nsk.su/archive/`
    `eaimage.asp?lang=1&did=6347&fileid=90637`
15. Notes received during the discussion on artificial intelligence (March 20, 1978),
    `http://ershov.iis.nsk.su/archive/eaindex.asp?lang=1&did=5147`
16. Ershov, A.P.: Diary entry (December 9, 1984),
    `http://ershov.iis.nsk.su/archive/eaimage.asp?fileid=161701`
17. Letter from Ershov, A. to Generalov, A. (June 27, 1988),
    `http://ershov.iis.nsk.su/archive/eaimage.asp?lang=1&did=7462`
    `&fileid=90536`
18. Notes of lectures on the differential variable theory (April 1955),
    `http://ershov.iis.nsk.su/archive/eaindex.asp?lang=1&did=12576`
19. Programme of the Royal Philarmonic Orchestra Concert,
    `http://ershov.iis.nsk.su/archive/eaindex.asp?lang=1&did=6368`
20. Happy birthday: sheet music and lyrics (December 1958),
    `http://ershov.iis.nsk.su/archive/eaindex.asp?lang=1&did=6363`
21. A.P. Ershov's Archive. Folders 267, 268
22. Ershov, A.: In-store Ideas and Wishes,
    `http://ershov.iis.nsk.su/archive/eaimage.asp?fileid=137023`

The START Project

Alexander Gurievich Marchuk

A.P. Ershov Institute of Informatics Systems of SB RAS
Acad. Lavrentiev Ave., 6, Novosibirsk 630090
mag@iis.nsk.su

Abstract. The paper overview one of the projects at the middle 80-th, supported by USSR government and oriented on breakthrough in hardware/software development. Several academic groups with industrial companies fulfilled a lot of investigations, built prototypes of perspective computers and intellectual programs. Author, as one of participants and headers of START project, analyses some principles of the project organizing structure. Although most brilliant results of START project were not used by industry because of economic and political crisis in Soviet Union, the impact of enthusiastic work and generating new ideas, was great for all major developers. Several opinions of participants are presented in conclusion.

Keywords: START, Modular asynchronous evolutionary systems, Kronos workstation, development of computing systems, hardware development, software development.

1 The Beginning

The temporary scientific and technical group START was created in 1985 by a joint decision of the Academy of Sciences and the State Commission on Science and Technology for a period of three years. The resolution [1] holds: "For the purpose of development and experimental testing of the elements of the fifth-generation computer concept..."

The resolution was preceded by the launch of Japan's national project of the creation of the fifth-generation computers in early 1980s. The land of the rising sun, having achieved substantial positions in the world automobile and home appliance industries, would not tolerate its meek results in computer engineering. Best minds and immense funds were directed at the solution of this breakthrough task.

The State Commission on Science and Technology wanted to find out how serious a threat was hidden in the Japanese project. The formed a working group headed by V. Kotov, the SB AS Computing Center deputy director at the time. The group included leading specialists of the Academy in the fields of computing engineering, software and artificial intelligence systems, as well as representatives of the industry. The group had a task – not only to evaluate the scientific advisability and technological feasibility of the Japanese project, but also to propose a concept of developing computers and software in the USSR.

It was a unique chance. The thing is that after the emergence of first computers, in which specialists from academic institutes participated immediately (with a creative

J. Impagliazzo and E. Proydakov (Eds.): SoRuCom 2006, IFIP AICT 357, pp. 126–133, 2011.

peak in mid-sixties in the form of the BESM-6 computer), the Academy of Sciences was "unchurched" from computer engineering; occasional teams were busy with innovations in programming. The result was the country getting its ES and SM computers that did not make a single step forward, but on the contrary, they cemented our technology lag. The new developments such as Elbrus and PS-2000 already did not seem revolutionary enough. The working group, which included such acknowledged specialists as E.H. Tyugu, Yu.G. Evtushenko, V.E. Kotov, A.S. Nariniani, V.M. Bryabin, and young but feathered researchers and developers who gathered from different towns and research groups of the USSR, took up the research enthusiastically. Due to security restrictions we had to work in complete isolation for a month; the result was an extended concept report that unfortunately was classified secret.

I remember well that there was no serious disagreement, not to say conflicts. There was a creative atmosphere, scientific analysis, and respect for the opinions of each other. Our report said very little of the Japanese project – we did not even want to criticize its flaws. The future has shown our rightness, and that the Japanese "wonder" had little practical effect. By the way, leading Western countries started their own national programs in response to the Japanese one, and their results turned out far more important. Our enthusiasm and creativity were fueled by the sensation of the beginning of large-scale, useful work.

The principles and outlines of solutions in our concept showed a direction (we concentrated on just one), which would later be called IT – information technologies. Without getting too deep into the technical details, let's say that we proposed a scheme of building user-oriented computers and software. On the top of the then-virtual structure rested the "brand-new" intellectualization tools.

Fig. 1. Alexander Marchuk with colleagues

I must remark that the core of the working group consisted of representatives of the Novosibirsk Computing Center: V.E. Kotov, A.S. Narinyani, E.P. Kuznetsov, and A.G. Marchuk. What underlay this trust? The reason was the MARS project (Modul'nye Asinkhronnye Razvivaemye Sistemy – Modular asynchronous evolutionary systems), which started in the SB AS Computing Center in 1982, and in three years noticeable results had been achieved. The concept of MARS was acknowledged in the scientific circles, the design of MARS-M supercomputer was underway, business connections were established with industry organizations in Moscow, Severodonetsk, and Kiev. The report we prepared was presented, studied and received support in the State Commission on Science and Technology, Academy of Sciences and related ministries. The result was that we were proposed to form a temporary (for a period of three years) team of developers to implement the proposed concept – the START Temporary Science and Technology Group.

2 Work

The core of START was composed of teams headed by the main participants of the working group. It was not due to current situation; it was because during the concept development we realized that our views on the aims coincided and that we complemented each other well and could work together. The group was substantially decentralized. The largest part of the group was located in Novosibirsk; it included specialists from the SB AS Computing Center, developers and designers from the Novosibirsk Branch of the Institute of Precise Mechanics and Computing Engineering, Special Design Bureau of Computing Engineering (now Design and Technology Institute of Computing Engineering), Independent Applied Laboratory of the Ministry of Instrument Making. In Moscow, the group included scientists and developers from the AS Computing Center and was headed by Yu.G. Evtushenko (currently an academician and director of the RAS Computing Center), and V.M. Bryabin. The Tallinn group united specialists from the Estonian AS Institute of Cybernetics and Special Design Bureau. Later, START was expanded by means of the Kiev "Microprocessor" Production Company. When necessary, specialists from other cities and organizations were invited. In different points in time, START included from 200 to 300 people.

The task was very specific: to develop and create a set of hardware tools and software defined by requirements specification. It meant pioneering developments done at the highest up-to-date international level. Such a problem could be solved in such short time only by an energetic, enthusiastic young team. We had it all in full volume. Verdant youth was involved both in research and research management; their opinion was always respected and considered. I, thirty-three at the time and in the position of vice-leader of the project, already felt like an old wizard.

The energy and creative spirit of the youth was fully expressed in the Kronos team. Kronos is the name of a 32-digit computer developed by NSU students, one of the first microcomputers in its class. Kronos is a legend which began before START. Kronos remains a legend, with a hint of nostalgia over youth, for all those who were involved in this extraordinary event.

Perhaps the story began at the Young Technicians Club of Akademgorodok, where Zhenia Tarasov and Volodya Filippov worked in different laboratories on a processor and a video display, respectively. In Kazakhstan Republic Dima Kuznetsov learned programming from books, because there were no computers in their town then. Or perhaps, it all started in the Physics Department dormitory of the NSU, where Zhenia and Volodia united their creativity; then teamed with Volodya Vasekin and "real" programmers from the Mathematics department – Dima (a sophomore) and Alexei

Nedorya (a junior); together, they started a club called *Intruder*. Figure 2 shows some of these technicians.

For me – and the story I'm telling – it all began in 1983, when the mentioned characters, at the time juniors of the Novosibirsk State University – only Alexei was a senior – came to me and declared: "We want to create a 32-digit computer! We know how to do that." I had nothing left but to support them – the kids were just too good. Youthful boldness combined with the fundamental knowledge NSU provides makes wonders if you manage to see in young people talent, enthusiasm, and literacy, and trust and help them afterwards. Wonders did happen – Zhenya defended the second (!) version of the Kronos processor as his graduation thesis, and Dima presented a cross-compilation system for this computer. By the way, Kronos-2 was replicated in as many as a hundred copies, because many wanted to have a "real" 32-digit computer and it had become available. Soldering and tuning was the duty of Volodya Filippov – and he still enjoys soldering, even though he has become the deputy director of the IIS.

Fig. 2. The Kronos Group: Dmitry Kuznetsov, Alexei Nedorya, Yevgeni Tarasov

An important peculiarity of the Kronos project was its exceptional attractiveness for talented students. I have never seen another young team so efficient and self-organized. On the forty square meters assigned for the project, something was always being programmed and discussed day and night, more and more new students joined the team and were assigned tasks, and in little time they became veterans of the Kronos movement. About thirty people took part in the project. I must say that all of them achieved high professional level and now lead their own teams – in different organizations, cities, and countries.

Returning to the START project, it was important to organize the work properly and achieve its efficacy. Modern generations of programmers can hardly imagine the limited conditions in which programs were developed in the 1970s and the 1980s. To improve substantially the instrumental support of the development it was decided to buy foreign computing machines. Not all remember that at the time the USA cast a strict embargo on sales of modern machines, including computers, to the USSR.

Hence, Australian-made personal computers and workstations were bought. Despite its disadvantages, the Australian computers were instrumental in solving the problems, since they allowed working (at least in programming) on a quite high technical level. I will not list the things we lacked, compared to our foreign colleagues, because even the capacities of "elite" machines were very limited; this lack was well compensated by the enthusiasm and professionalism of the developers.

The work was organized by team contract method, i.e. a team took a certain task under collective responsibility. Hence, the relationships within the teams were very democratic, and the team leader performed mostly provisional and external coordinating role, rather than keeping to regular directorial duties.

Key elements in the START activities were the annual January meetings in Ivanteevka. Ivanteevka is a small industrial city near Moscow, where START activists gathered in summer house areas for mutual reporting, planning and coordination. Why January? Perhaps because our first meeting took place before the official start of the project, when we formed and defined exactly the requirements specification, which served as basis for official documents that were being prepared for the first of April. It later became a tradition.

"Ivanteevka", as we called our meeting, was a one-of-a-kind conference. Machinery was brought – computers, working samples, an overhead projector (a true novelty at the time!), the work was organized in such a way that we could really discuss, report and immediately make changes to our results, develop prototypes. This was a real two-week creative workshop, where everyone made his valuable contribution to the discussion and realization of general and special issues. We never had time to sleep; luckily, we all were young and full of energy. A huge project never develops linearly, there is always a need for substantial corrections, and those corrections were made, having been thoroughly discussed every time. For example, it was intended initially that a product of the Tallinn design bureau, made from standard components, be used as the basis of the Kronos. However, after the high results of joint evaluation of the students' Kronos project it was decided to reengineer Kronos according to the project and industry specifications and use it as the basis for workstations produced in Tallinn and a model of super-mini-PC that was being developed in Novosibirsk.

In addition to other things, Ivanteevka facilitated personal friendship in our distributed team. This friendship goes on today, having turned into a kind of "fraternity". No matter where you go, if there are ex-START people, you will be infinitely welcome, and there will be no end to warm memories.

START did not go unnoticed by the press and television. Their attention was, however, quite special – seeing a "wonder" and telling about a wonder, but not about good, wisely organized professional work. Repeatedly we asked the reporters to show us their texts at least for minor technical examination – and not a single time they considered it necessary. They would write all sorts of nonsense... Official authorities also liked to use our materials for solving their own problems, but things said from the tribunes reflected the state of affairs very roughly. For example, Mikhail Gorbachev, the Communist Party General Secretary of the time, told at a regular meeting of VLKSM (Leninist Young Communist League of the Soviet Union), that a team of researchers headed by a young Dr. Kotov (it's hard to believe the young communists would think of a fifty-year-old as a young man) created a supercomputer capable of performing 100 million operations per second. We did not have such tasks set before us.

Yet we could not find solutions for the social problems young people inevitably face. The housing provision system was so crisis-ridden that every single place in a dormitory or an apartment cost us a grand battle and did not at all improve the overall situation. The lasting suspense of domestic problems served as one of the substantial reasons of the massive flight of our young talents abroad.

3 Results

Exactly at the target date, at the end of the three years, a government resolution created the State interdepartmental commission for the acceptance of work completed by START. We had things to show.

The following is a partial list of results presented to the commission.

- A four-processor working model of MARS-T multiprocessor complex, built on the basis of the developed processors Kronos 2.6, equipped with a multiprocessor multi-program operating system that we developed, which successfully completed the given test parallel programs;
- The PIRS workstation equipped with a full set of system software, including two operating systems, Unix and Excelsior (a product of the Kronos group), supplied with a software package sufficient for numerous applications;
- BARS parallel programming language, Polar architecture design language, and NUT programming system, providing new intellectualization tools;
- A set of programs for Specter workstations and personal computers;
- Modula-2, C++ and Fortran compilers;
- VLSI CAD system; graphic tools, engineering CAD system – the list continues.

The first two years of START showed that something brand new, worthy of replication and application in the national economy was being developed. That's why on the final stage we invited fresh teams of constructors, industrial software designers and microprocessor designers. It was decided to aim at the future prospects, involving into the introduction process special institutes, design bureaus, and plants. In the result, the completion of START didn't mean the end of works, because research had turned into development and creation of special systems. Scientific and creative contacts multiplied and spread over cities and countries, primarily within the COMECON countries. We planned that:

- Kronos be installed as on-board computer of airplanes and space vehicles, and the development of the microprocessor set went into its final phase;
- The production of workstations and parallel computers attracted the attention of all sorts of businesses;
- Software was introduced and multiplied on all kinds of platforms, including the then-new PCs;
- CAD systems were adapted to rigid conditions of industrial design.

Unfortunately, many of our intentions never found any development due to the economical and political collapse of the late eighties. Nevertheless, a noticeable proportion of our products found its place in the new conditions. Analyzing our results, I see the most important achievement not in the final product, but in scientific results and creative professional growth of the people for whom START was not only work, but also a time of steadfast headway, a time of establishing oneself as a specialist and leader. Life scattered STARTers all over the world, but most of us retained our loyalty to professional duty and calling. Our relationship remains friendly and warm. Science is an international business, and both those of us who happen to work abroad and those who work in the Russian IT industry do the same thing, recalling our youth as an example of fine work.

4   Opinions of Participants about START

Enn Haraldovich Tyugu, Institute of Cybernetics, Tallinn, Estonia: "The NUT programming environment http://cs.ioc.ee/~nut/ was gradually tuned up to the state of ready product. Its application yielded a number of interesting scientific results, and it is still used by researchers for modeling complex dynamic processes" [3].

Vitaly Telerman, currently a leading developer in Dassault Systèmes, France: "On the one hand you can say that the results achieved in START 15 years ago have finally been implemented into the industry, somewhat "French-style", perhaps. Unfortunately, not everything has been done, some things are only starting to be taken seriously by developers; much remains unused in practice". [2]

Viktor Mikhailovich Briabrin, Broad Street Software Group, Мериленд, США: "In the days of START, everything was devised and implemented on the spot with creativity, skill, elegance, and enthusiasm thanks to the talent and professionalism of our fellow workers and postgraduates - participants of the project. Of course, it can be said about most of the START project members – in Moscow, Novosibirsk, and Tallinn". [4]

Vadim Evgenievich Kotov, Carnegie-Mellon University, USA: "After some creative re-thinking, I pursued the ideas of modular asynchronous evolutionary systems, which served as a basis for START, in my work at Hewlett-Packard Laboratories on Systems of Systems architecture. These works led to the emergence of Service-Oriented Architecture, in which all levels from top to bottom are federations of independent interacting components – services located in a network-federation of independent servers. I use the same approach in our current works on the creation of high-reliability systems for NASA". [5]

Alexander Semyonovich Narinyani, Russian Scientific Research Institute of Artificial Intelligence, Moscow: "our greatest adventure was the summer that we spent in Gury Ivanovich Marchuk's office, developing the project concept and calling up a whole crowd of experts and administration officials for the discussion of different aspects of this challenging document". [6]

Andrei Khapugin, Excelsior, Ltd.: "Most often I recall an article from, as far as I remember, Komsomolskaya Pravda about the Kronos group, which started with words: "Dmitry Kuznetsov at 25 is a super-programmer of the international level, and he is completely aware of that". They didn't give him a minute of peace for weeks afterward...". [7]

Acknowledgment. This paper is essentially based on an article published in the Nauka v Sibiri (Science in Siberia) newspaper [8].

References

1. Resolution of State Committee on Science and Technology and Presidium of the USSR Academy of Sciences (March 19, 1985), 101/48
   http://start.iis.nsk.su/archive/eaindex.asp?lang=1&did=23849
2. Telerman, V.: Recollections about START,
   http://start.iis.nsk.su/persons/telerman/index.shtml
3. Tyugu, E.: START in Tallinn and all that ensued,
   http://start.iis.nsk.su/persons/tyugu/index.shtml
4. Briabrin, V.M.: SPEKTR Project,
   http://start.iis.nsk.su/persons/bryabrin/index.shtml
5. Kotov, V.E.: Answers to the questionnaire,
   http://start.iis.nsk.su/persons/questionnaire/kotov.shtml
6. Narinyani, A.S.: Answers to the questionnaire,
   http://start.iis.nsk.su/persons/questionnaire/narinyani.shtml
7. Khapugin, A.: Answers to the questionnaire,
   http://start.iis.nsk.su/persons/questionnaire/khapugin.shtml
8. Marchuk, A.G.: 20th Anniversary of Legendary START Project, Nauka v Sibiri, no. 16 (2502) (April 2005)

The MRAMOR Workstation

A.A. Baehrs

A.P. Ershov Institute of Informatics Systems, SB RAS
baehrs@iis.nsk.su

Abstract. The paper describes the experience of creating the MRAMOR workstation in 1980-1987, a quality workspace for the publishing business built on a weak element basis. The aim of the work consisted in the creation of hard-and-software foundation and a system of workspaces for professional publishers. This accounted for the initially complex approach to the problem, which combined hardware, software, font, and visual design. We produced a pilot batch of forty workspaces, basic program software for the station, and application software for workspaces of professional publishing systems for electronic publication preparation. These convenient and highly effective workspaces went into test operation and they served to produce a large number of publications of high polygraph quality.

Keywords: Workstation, electronic publishing, parallel complex development of software and hardware.

1 Introduction

This paper dwells upon the experience of the creation in 1980-1987 of the MRAMOR workstation, a quality workspace for the publishing business built on a weak element basis. The project, dubbed "The RUBIN (RUBY) project for the PRAVDA newspaper", was part of international cooperation between the USSR and the People's Republic of Poland. The Soviet side was represented by the Pravda publishing office, the SB AS USSR Computing Center, and the M.V. Keldysh Institue of Applied Mathematics, while the MERA-Blone precision mechanics factory and the COBRESPU—Center of television equipment—represented Poland. The RUBIN project was part of the list of key research projects of the GKNT USSR (Gosudarstvennyj Komitet po Nauke i Technike, USSR State Commission on Science and Technology) in the years 1980-1985.

An academic institute developed the architecture, design, and functional characteristics of the MRAMOR workstation and of the MRAMOR-based workspaces, as well as its basic and application software. The design group at the MERA-Blone factory performed the realization of the developed ideas into working machinery and for preparation for serial production. The Pravda publishing office substantially supported the work within the RUBIN project.

The developers' collective consisted of the following teams:

The Laboratory of Experimental Informatics of the Computing Center of SB AS USSR was represented in the MRAMOR WS project by A.A. Baehrs (GEO), Yu.

J. Impagliazzo and E. Proydakov (Eds.): SoRuCom 2006, IFIP AICT 357, pp. 134–141, 2011.

Bovkun, A. Kovalenin, A. Melnik, A. Mullagaliev, G. Nesgovorova, E. Ovcharenko, V. Polyakov, S. Rudnev, M. Sadomskaya and V. Chetvernin.

Our polish colleagues at the MERA-Blone factory were M. Augustinyak, J. Zawadsky (SEO), J. Zagraek, A. Kolodeyak, Z. Luchuk, J. Matrash, T. Moshevich, R. Pacek and S. Shumsky. The Warsaw Center of television equipment (COBRESPU) under the direction of J. Kania and L. Nepekla developed the video displays for the MRAMOR WS. All current organizational tasks of the RUBIN were the job of the technical department of Pravda publishers, and a personal duty of their leader, V.A. Tiefenbach.

The author, who was the general executive officer and chief designer of the project, considers it a pleasant duty to mark the exceptionally friendly and creative interaction of all members of this international collective, consisting of specialists from many fields. I am still cordially thankful to all the members of this interesting and long work. Additionally, I wish to acknowledge the care and attention the project received from our teacher, academician Andrei Petrovich Ershov.

2 The Project

The stages of production process research in the Pravda publishing office and of system analysis of these processes preceded the start of the RUBIN. The results of the research, summed up in the "General Scheme of Creation and Development of the RUBIN system for the PRAVDA Newspaper" and approved by the Pravda editorial board in February of 1979, held that the system would consist of a peripheral network and a central computational complex (CCC). The CCC, with a large information and reference database for analysis and verification of published materials and perspective planning of newspaper issues, appeared on the older models of ES computers. We created the MRAMOR workstation as the terminal base of the peripheral local network of workspaces for the publishing office employees. It was necessary to automate laborious editing and publishing processes to make it possible to process large volumes of textual information in strictly limited time, which is of crucial importance in newspaper publishing.

The aim of the work consisted in the creation of hard-and-software foundation and a system of workspaces for professional publishers. This accounted for the initially complex approach to the problem, which combined hardware, software, font, and visual design.

We produced a pilot batch of forty workspaces, basic program software for the station, and application software for workspaces of professional polygraphic systems for electronic publication preparation. These convenient and highly effective workspaces went into test operation and they served to produce a large number of publications of high polygraph quality. These workspaces included regular publication of the ENSK city newspaper, preparation of a number of issues of Nauka and EKO magazines, and a large number of books.

Serial instances of MRAMOR were shown at the "Siberian Device 87" exhibition (Akademgorodok, 1987), where the design received the Diploma for the original solution, at the All-Poland POLIKON-87 Conference (Poznan, 1987), and at "MERA-Blone Factory – to Soviet Informatics" exhibitions in Moscow, 1987, and Vilnus,

1988. At the 1988 SB AS USSR projects contest, the MRAMOR received the Diploma of Third Degree award.

3 Main Features

The inspection of the Pravda publishing and printing offices in 1977, the results of which were presented in the report called "System analysis of production processes of the publishing of Pravda newspaper" allowed to highlight the main channels of information processing and their substantial characteristics necessary for the creation of RUBIN. These are as follows.

o   processing of texts, typesetting, make-up and preparation of newspaper typing range print forms
o   planning and management of issues and publication
o   information and reference service and proofing of published facts

Various workspaces for editors, publishing and printing office employees must provide cooperative coordinated access to these three streams, realization in various combinations (defined by the workspace type) of the needed applied functions, and a possibility to use results of one work in another.

3.1 Considerations

It was necessary to take into consideration several mandatory limitations stipulated by the customer and situation during the realization of the RUBIN. These included the fact that we had to build the equipment chiefly on domestically made basic elements and components. As an exception, they allowed us to use microchips, components, and devices produced in COMECON countries. This understanding was essential to retain maximally the technology of work that had been established in the editorial office and the style an interaction principles of its employees. There was an understanding that users of such social rank would disagree to adapt to inconvenient functions and features of equipment and programs.

On the other hand, one should remember that by the end of the 1970s, the computer stock in the USSR consisted mostly of ES and SM computers equipped predominantly with borrowed, so-called "States" software. They were totally unsuitable for work in the editing business. Terminal equipment acceptable for editing and publishing in preparation of publication was practically absent; the domestic Kaskad phototypesetting automata complex was morally outdated; font supplies of the phototypesetters and terminals were plainly primitive. In essence, the foreign polygraph computer systems employed at the time were poorly suited to the needs of local practice and created difficulties in publication in Russian and languages of other peoples of the USSR.

3.2 Construction

It was possible to make the CCC from one of the older ES models to create application software using the existing OS and DBMS. However, it was obvious that for a terminal system of the RUBIN level, it was impossible to solve automated workspace problems only by means of writing software for any of the existing and practically available mini- or microcomputers. Thus, we decided to create a new machine of the macro-mini class, dubbed "the MRAMOR Workstation".

Considering the available element base and form factors, it was clear that the computational powers of a single K580VM80 microprocessor would not suffice to provide the needed functionality and features of workspaces. Hence, we solved the problem we faced by parallel development of software and hardware for the new complex. We also had to keep in mind the practical limitations of the so-hardly-found production plant, the polish MERA-Blone precision mechanics factory.

3.3 Requirements

The development project of MRAMOR included the following requirements to the functions of the new hard-and-software complex:

- o modularity of hardware, basic and application software;
- o multi-processor architecture, including the possible heterogeneity of hardware and software;
- o multi-tasking in serving the user with multi-program task load of processors;
- o multi-seating of the workstation with connection between workspaces;
- o multi-window interface, including the multi-screen feature of the workspace;
- o federativity, i.e. operational interaction allowing users, processes and processors to work in group mode
- o high quality of representation on the screen and in working printouts of polygraph aspects of the processed publications, including the multitude of national languages, writing types and fonts.

For all these various requirements to be realized within a common approach that would unite the possibilities of hardware and software tools and allow to make decisions on the project, a conceptual model of computational process execution organization was developed. Moreover, the peculiarities of preparation of different publications in the Pravda publishing office and the experience gained during the development of SAPFIR system for the *first* model-printing house provided a conceptual model of supporting the electronic preparation of publications.

These conceptual models, together with the above listed requirements, became a basis of the MRAMOR WS development, which lead to solving a number of all-new problems. These are as follows.

- o Practical realization of parallel design of united whole hard-and-software environment, including equipment, basic and application software, user interface, and workspace design;
- o Development of original open equipment, with heterogeneous computing blocks, and provide the necessary characteristics of workspaces even despite the weakness of the available elements, through the compensation of this weakness using of architectural and engineering solutions by means of possible choice between hardware and software realization of the needed functions;

o Creation of computations organization model – high-level operational environments – allowing to organize parallel execution in the form of working mixture of a large number of interacting processes flowing for the purposes of several users on a multitude of virtual (including real) processors, each of which works in multi-program mode;

o Development of an operational system and software controlling the work of the multiprocessor complex on the basis of the given computational model;

o Creation of a multiprocessor system from heterogeneous components and organization of their joint work in the mode of multi-program execution of a large number of linked processes for the purposes of several users simultaneously, which would provide the necessary degree of parallelism by organizing a working mixture of virtual machines on heterogeneous computing tools.

o Creation of a possibility of further extension of MRAMOR WS functions through addition of modules, devices, and software, resulting in a wide range of workspace configurations available with MRAMOR, using a small number of produced module types.

In this way, MRAMOR embodies a number of original architectural, engineering and program solutions, some of which have become ubiquitous (multi-bus architecture with heterogeneous processors, portrait display layout, multi-window interface, chord keyboard input, and programmable fonts with different tracing of variable-width symbols).

3.4 Architecture

The pilot batch included two MRAMOR WS models: a two-seater and a four-seater. In addition, we produced two instances of single-seater multi-screen workstation for the needs of the developers. The computing blocks of the station develop around the 8086 processor and its coprocessors: arithmetic and input-output. The block links the system and local buses and contains schemes of 15 termination inputs, where we could connect any of them to any of the termination sources, the mechanism of page addressing, and EROM microchips with volume totaling 32K of 16-bit words.

We developed an original mechanism of addressing extension for the computing blocks: a scheme that defines the purpose of processor calls to the memory by state signals and by the ordinal number of cycle. The 16-bit address of command access, stack, or data processing was supplied with one of the three-order program-defined prefixes, stored in a special register. Thus, we split the main memory into four subspaces, any three of which one could simultaneously assign to the program as subspaces of commands, data, or stack.

The system provided page access to memory using hardware-implemented re-addressing table (RAT). The processor could read or write directly any word in RAT. There are two page size variants – 4 Kbytes and 16 Kbytes. The RAT generates signals directing the address either into the permanent memory or into its local bus or system bus (and further through the system bus into local buses of other blocks). In the 16 Kb variant, it can use all bits of processor address, the 14 lower ones defining the page shift.

In the result, the processor has simultaneous access to 64 pages, each of which can be in any of the independent address memory subspaces: the system bus, the local bus of any of the computing blocks, and permanent memory of the given block. The logical range of the system bus addressing is 16M, of local bus, it is 1M, and of permanent memory, it is 64K. Access of each of the processors through the system bus to a local bus of other computing blocks has priority over access from the local processor. Hence, if the processor sends a "lock signal" to the system bus, this blocks access to it from other computing blocks. The correctness of work with commands of the test-and-set type is retained even when they are placed in the local memory.

The ONIKS operating system allowed using several different type processors in "teamwork" mode and provided for "mixed execution" of working mixture for several languages – Yava (Yazyk Vmesto Assemblera, Language instead of Assembler), Cidula (C plus Modula), and Fodula (Forth plus Modula). MRAMOR-based workspaces allowed developing software, creating fonts and preparing publication of a number of newspapers and books, including, for example, a Russian-Chinese phrase book.

3.5 Physical Considerations

In one of his last articles, the editors column in the magazine "Mikroprocessornye sredstva I sistemy, No 2, 1988" — *"To look ahead, but to see around"*, Andrei Ershov brought to light the direct parallels between an American forecast list of chief PC characteristics in the nineties by W. Beem and the features implemented in the MRAMOR workstation. In the process of creating MRAMOR, we paid much attention to the general layout and design of workspaces. Unlike the universally accepted approach, when one places a PC, its display, and other blocks on a desk and occupying most of its surface, the designers of MRAMOR kept in mind the preference for a "large, spacious working desk".

That is the reasons all modules and the FDD are placed in a wheeled side-table; the keyboard is set on a pull-out frame located underneath the desk surface and allowing configuring position and angles. The keyboard had 137 buttons, 134 of them fully pre-programmable; that is, a driver defines their layout and semantics.

We arranged the buttons on the keyboard into six ergonomic fields, three of which are functional button fields; one corresponds to the typewriter keyboard and the other two for numeric input and cursor manipulation. The five register keys enable working in 32 registers with 128 symbols in each, which allows using as much as 4096 symbols in a single publication. Button fields are color-coded; intervals in which one can put stickers with function captions separated them. There are eight signaling

LEDs of colors red, yellow, and green with three of them connected to fixed-position buttons.

MRAMOR contained blocks of direct output onto the phototypesetting part of the FA-1000 device and the Gazeta-2 photoreceiving device, which provided quality offset printing forms. MRAMOR also featured software power control because some workspaces could be located in other rooms. Upon finishing a session the user can turn off only his local workspace devices (display and FDD), but not the whole workstation. The basic software did the rest.

Talking of an editor's workspace, the key parameter defining its comfort and usability is the visual quality of textual information, allowing one to display a wide range of symbols in a variety of mark-ups. Hence, we supplied the MRAMOR with a custom-ordered monochrome 50-cm display with high persistence rate to eliminate flickering on the screen with standard TV scanning. Moreover, the declared the use of permanent memory for the generation of symbols as inflexible, and we chose a display scheme featuring support memory. For better correspondence of the image to the format of printing ranges, it was decided that the display be oriented vertically relative to its longer side (the so-called portrait layout, as opposite to the landscape layout seen in most TVs).

At the same time, we retained the standard TV scanning, i.e. vertical, across the text lines. The 64 Kb memory device of the image allowed creating a 768x576-dot field with three levels of brightness and we hung the display over a table on a special rotating holder enabling variable height and tilt.

4 The Publication Preparation System

The MRAMOR WS is a professional system for electronic preparation of polygraph publications of different types and complexity. It provided the publisher with a complete technological cycle from the initial text input to the make-up and production of ready photo forms. We based the system on the principle of separating the text of a publication from its polygraph design. The text itself appears as a double sequence. Its higher level consists of heterogeneous publication elements accessed through a working directory, which also contains types and external characteristics of the elements.

The polygraph design of the publication is coordinated with element types of the publication and/or text mark-up and represented by a separate range-by-range structure. We accomplished the intra-textual links and footnotes by markers, which are included in the mentioned structures and made the dialogue development of a mechanical and its range-by-range makeup according to the working directory; that is, a system of linked windows support editing and correction. We represented fonts by special display fonts; however, we calculated the line formats with real widths of typesetting fonts.

5 Conclusion

Our team developed technologies and tools for electronic preparation of publications by means of separating text from its polygraph realization, and possibilities for their separate processing. We developed this with the parallel use of structure-referential representation of the text body combined with text markups for efficient global processing of the text (including the making of mechanicals and make-up). We established the notation of workspaces in different subdivisions of the publishing house, together with the possibility of dynamic transition to functions of different logical workspaces from a single location in the process of work. In the end, our team provided the possibility to prepare a mixed publication in different languages with writing systems within a single workspace. We all felt that this achievement was a true accomplishment.

References

1. Baehrs, A.A.: Data-processing System RUBIN for "Pravda" newspaper. Applied Approaches in Informatics, 55–78 (1980) (in Russian)
2. Baehrs, A.A.: Software for Typeface Display. Experimental Informatics, 51–80 (1981) (in Russian)
3. Baehrs, A.A., Polyakov, V.G., Rudnev, S.B.: On high-level programming system with mixed computation for personal microprocessor systems. Topical Issues of Computer Architecture and Software, 78–94 (1983) (in Russian)
4. Baehrs, A.A., Polyakov, V.G.: Multifunction workstation architecture for editors' office. Personal Computers in Informatics Problems, 40–49 (1984) (in Russian)
5. Baehrs, A.A., Polyakov, V.G.: Look-and-feel of the system software for editors' office multifunction workstation. Personal Computers in Informatics Problems, 50–57 (1984) (in Russian)
6. Baehrs, A.A.: New-generation workstation MRAMOR. New-Generation Computers Design: Architecture, Software, Intellectualization, 126–141 (1986) (in Russian)
7. Baehrs, A.A.: On object orientation and architecture organization of software systems. Topical Issues of Programming Technology, 4–15 (1989) (in Russian)
8. System analysis of production processes of "Pravda" newspaper publishing. Joint report of Computing Center of SB AS USSR and NB IPMCE, Novosibirsk (1977) (in Russian)
9. Baehrs, A.A.: General scheme of creation and development of data-processing system RUBIN for Pravda newspaper. Pravda Publishing House, Moscow (1979) (in Russian)

Full texts of the documents [8, 9] stored in A.P. Ershov archive are available at http://ershov.iis.nsk.su/ (Software projects section: MRAMOR, RUBIN).

Mixed Computation in Novosibirsk

Mikhail Bulyonkov

Ershov Institute of Informatics Systems of the Siberian Branch of the Russian
Academy of Sciences
630090, Novosibirsk, pr. Acad. Lavrentieva, 6
mike@iis.nsk.su

Abstract. In these notes, I would like to give an account of the history of mixed computations at Novosibirsk Computing Center and later at A.P. Ershov Institute of Informatics Systems. It is quite possible (and even most probable) that I will not be able to mention all persons and events relevant to the works in this field, but in no way, it is to diminish their significance.

Keywords: Mixed computation, partial evaluation, program specialization.

1 Futamura Projections — Andrei Ershov

Andrei Petrovich Ershov came to the idea of mixed computation from his work in compiler construction, program optimization, and transformation. In his first works on the subject, mixed computation processor, or *partial evaluator* was defined as program processor that has as its input some program representation and part of its input data, and produces as its output transformed program, partial result, and the data demanding further processing.

Next thing Ershov noticed was the fact that compiler's behavior is very similar to that of mixed computation: operations of the compiled program analysis alternate with object code generation. These considerations culminated in discovery of relationships between program interpretation and its translation into object code. For this purpose, they interpreted a partial evaluator in a more restricted sense – without intermediate data and partial result:

$$mix(p, x) = p_x$$

such that

$$p(x, y) = p_x(y)$$

Let us call the last equality a *mixed computation equation*. Let *int* be interpreter of some language L, written in the language whose programs can be processed by *mix*. Then

$$int(p, d) = p(d)$$

for each program p in L and its data d. Then, substituting *int* in the mixed computation equation, we get

$$int(p, d) = int_p(d)$$

J. Impagliazzo and E. Proydakov (Eds.): SoRuCom 2006, IFIP AICT 357, pp. 142–151, 2011.

whence

$$int_p(d) = p(d)$$

Thus, programs int_p and p are equivalent, but note that we write them in different languages: p in L, and int_p in the interpreter language. In other words, mixed computation applied to interpreter implements compilation, and int_p is an object code for p.

We can obtain very interesting relations under the assumption that partial evaluator possesses the property of *self-applicability*. For this to occur, we must write it in the same language as the one for which it is intended. We leave to the reader the proof of the following relations:

$$mix_{int}(p) = int_p$$

and

$$mix_{mix}(int) = mix_{int}$$

Thus, mix_{int} transforms a program into object code and hence implements the function of compiler, while mix_{mix} transforms the language semantics, defined in the form of interpreter, into compiler and therefore appears to be compilers generator.

Anybody who had a faintest notion of compilers and interpreters and who became familiar with these relations, at first could not believe in them and looked for some hanky-panky trick. They then became delighted with the beauty, clarity, and depth of their meaning[1]. Today I can clearly imagine Ershov's disappointment at having discovered a paper by Japanese researcher Yoshihiko Futamura, who had published these relations back in 1971. Andrei Petrovich called the independently found relations Futamura projections, although the last of the three was not present in Futamura's paper; it serves as an excellent example of professional ethics.

The main approaches to implementing mixed computations appeared as early as the first works on the subject. They include:

1. *Partial evaluation* refers to the immediate execution of all instructions from the source program that depend only on available data. We reduced the remaining instructions and placed them into a residual program.
2. *Generating extension* splits the process of residual program construction into two stages. First, we classify all actions of a program as "available" or "delayed". Then we generate a program that, given as input the available data of the source program, becomes the residual program. People recognized the importance of the static classification of a program's actions as available and delayed much later; they called it *binding-time analysis*, or BTA.
3. *Transformational approach* refers to obtaining the residual program from the source by application of a sequence of so-called reducing transformations such as constant propagation, reduction of expressions and conditionals with constant conditions, loop unfolding, and the elimination of unused computations.

---

[1] I heard that Andrei Petrovich made a long-distant call to his son Vasily and for several hours described his new findings. Similar story is told about Valentin Turchin, who also found these relations independently.

We recognized rather quickly the main problems of mixed computation – a problem of delayed control and related problem of mixed computation termination. In his first works on mixed computation, Ershov neatly disguised these problems, which he acknowledged in a reply to my pointed questions. It had taken years of research and experiments to arrive to a practical implementation from a sketchy idea. However, back then the main goal was to introduce the computer science community to the idea of mixed computation potential and Ershov dealt with it brilliantly. Dozens of times, at various conferences and seminars, in papers and informal talks he spoke about Futamura projections and mixed computation, forming a new school in system programming. His large comprehensive article with many color illustrations that appeared in the popular-scientific journal "V mire nauki"[2] [Ersh84] serves as spectacular example of this activity. The works in the mixed computation field brought Andrei Ershov the prestigious Krylov Prize of the USSR Academy of Sciences.

2   Algebra of Mixed Computation — Vladimir Itkin

A rather different background and approach to the study of mixed computation appears in the works of Vladimir Itkin. He gave an impression of a withdrawn, cloistered scientist who valued strictness of proof and theoretical power of result much more that its practical applicability — a typical example of a "theorist"[3]. He wrote several of his first papers in co-authorship with Ershov, but subsequently their co-operation became less close. The motive was Ershov's words that in their joint papers, the general idea was his, while Itkin worked over the technical details. Even subsequent clarifications — *only* the general idea and *all* technical details — apparently were not enough for Vladimir. Although he worked independently, he obtained results and gave a solid theoretical and conceptual basis to other researchers.

The starting point of Itkin's research was a theory of program schemata, oriented in the first place to imperative operator programs over common memory. The most actively used notion was that of explicator — an operator ensuring the required state of some part of program memory. In the simplest case, an explicator is expressed by assignment of given values to variables.

The role of explicators was twofold: on the one hand, being ordinary operators, they were part of the program, so at any moment, the mixed computation process could terminate and the current program declared the residual one. On the other hand, explicators were like current computation points. Vladimir Itkin analyzed and substantiated different patterns of mixed computation process (e.g. end-to-end, dotted, polyvariant). A substantial part of the process was the manipulation with explicators such as the "application" of program statements and merging. Itkin managed to abstract these operations by introducing the set of axioms they have to satisfy. It has lead to the so-called *algebra of mixed computations* [Itkin88].

---

[2] The Russian edition of *Scientific American* journal.

[3] I was always amazed by Itkin's "cut and paste" technique of paper writing. If he had to insert or reposition a couple of sentences in his written text, he did it not by editor's markup or by notes on the underside, but literally cut and pasted a piece of sheet with the text to the right place. The result of his effort amounted to a long paper belt picturesquely snaking through his whole office. Upon completion of works, all that remained to be done was cut it into standard-sized pages and give to the secretary for typing.

Vladimir Itkin's last works were of a philosophical nature. Partial evaluation interested him as a fundamental process of a transition from the general to the special. He died tragically in 1991 when he froze to death on his way to church.

3 Parser Specialization — Boris Ostrovsky

To make the idea of mixed computation more viable, it was essential to find some practical application for it. Ershov proposed this issue to his post-graduate student Boris Ostrovsky as a subject of his Ph.D. thesis. His task was to transform automatically a universal parser for some class of grammars into specialized grammar-oriented parser [Ostr87]. It was clear that parser specialization, although being of independent value, was only the first step on the way to Futamura projections implementation, since we can consider a universal parser as an interpreter, a grammar as an interpreted program, and the input string as its data. From this point of view, only the first projection was at issue and the task of achieving partial evaluator's self-applicability was out of question.

Ostrovsky based his partial evaluator on the transformation approach. To be more exact, it was not a proper partial evaluator, but rather a transformation machine with an ad hoc set of transformations. The process of transformations application was similar to the process of Markov's normal algorithms execution. Transformation markers labeled the text of a universal parser. Usually the process started with a single marker of initial transformation placed before the first program instruction. A non-deterministic iteration of marker selection and application of its transformation followed to the construct labeled by it. The transformation resulted in a program modification and a placement of new markers. The process repeated until no marker remained.

For each universal parser one had to develop the set of transformations individually. In fact, there was a separate partial evaluator for each class of grammars. However, even this approach was technologically justified. First, the basic transformation set was reusable, which guaranteed the ad hoc parser correctness. Second, they could use a single partial evaluator, after being "tuned" once, for a number of grammars. Third, the system supported sophisticated tools for grammar transformation such as regular part extraction and post-processing tools applied to residual programs.

The mixed computation system developed by Ostrovsky was one of the most - if not the most - advanced for those days. Unfortunately, the routine of teaching at the remote university kept Boris from active continuation of the work in this field.

4 The Limanchik Summer School on Mixed Computation

In my opinion, research on mixed computation reached true national level after the conference on mixed computation held at the "Limanchik" summer camp of Rostov State University in 1983. The conference was an out-and-out success! It afforded the opportunity to meet and establish good professional contacts to all active researchers in this field. There is no point in discussing the whole conference program; instead, I will note only two memorable scenes.

Viktor Kasyanov gave a talk on the reducing program transformations. The set of transformations was wide enough and constituted a substantial part of the SOCRAT system aimed at processing FORTRAN programs. One of the system's features was the fact that transformations were based not only on a program's text, but also on user annotations. We considered these annotations as "a priori" true statements about a program that could both reduce complexity of necessary analysis and increase applicability of transformations. During the ensuing discussion, Nikolay Nepeivoda took floor and declared in his usual offhand manner that not a single program exists to which one could apply such transformations. Later, it turned out he meant to say that no programmer would deliberately write a branch instruction with an always-true condition. However, in the case of automatically generated programs, application of such transformations can be very promising.

In another discussion Sviatoslav Lavrov cast doubt on practicality of Futamura projections for production compilers, since they cover only the most simple and well-studied syntax-directed part of the process. The most interesting problems of compilation such as register allocation or common subexpressions elimination have origins in neither a partial evaluator nor an interpreter; therefore, they cannot appear in automatically obtained compiler[4].

5 Polyvariant Mixed Computation — Mikhail Bulyonkov

Another Ershov's post-graduate student, Tatiana Shaposhnikova, suggested the idea solving the problem of non-termination of interpreter specialization. When she told me about her idea for the first time, I was unprepared to perceive it, since Tatiana's suggestion led to an introduction of unstructured *goto* statements. In my opinion, it posed a problem, because she essentially based the existing mixed computation patterns on program structuring as in the propagation of available information through delayed conditionals.

I succeeded in finding sufficient conditions of specialization process termination. If the binding-time analysis classifies variables and, therefore, program instructions statically (i.e. if we declare a variable static, it remains so at any moment of specialization), and the range of each available variable is finite, then each branch of polyvariant specialization will inevitably lead to an already passed state. In certain sense, computation turns out to be "shut" inside a matrix with rows corresponding to initial program instructions and columns corresponding to available memory states. Each cell of this matrix contains the corresponding reduced instruction.

Andrei Petrovich grasped this idea with enthusiasm[5] – it has become the key to implementation of self-applicable partial evaluator Mix suitable for carrying out all three Futamura projections. To reduce "technical" problems, we designed a special imperative language IL whose set of basic operations included among others the

---

[4] N. Jones [Jones88] later formulated and researched a similar but more abstractly stated problem: "Can program specialization yield more than linear acceleration?" Here we suppose that complexity measurement takes into account only the delayed data size, and the available data size is constant.

[5] It was Ershov who translated into English (or to be honest — rewrote in English) my paper [Bul84] that was later extensively cited.

operation of expression reduction. We presented the methodological comprehension of the obtained results in our joint paper [BulErsh86].

Unfortunately, we were a little late again. Just a few months earlier, Peter Sestoft, under the supervision of Professor Neil Jones from DIKU, showed the successful implementation of all Futamura projections for a small subset of the functional language LISP. In turn, they were surprised to discover belatedly the paper [Bul84] that described the technique they independently found during their partial evaluator implementation. However, these small disappointments became the starting point of a long-term collaboration, mutual visits, and healthy competition between DIKU and Novosibirsk group.

6 Partial Evaluation and Compilation Phases — Guntis Barzdin

Guntis Barzdin came to Novosibirsk to negotiate the possibility of his post-graduate study at Latvian University with Ershov as his scientific adviser. After having settled his affairs, Guntis paid me a visit and was very surprised to find me being only a few years older than he was. He had read my papers and pictured me as reputable man of science that was contrary to fact. During our discussions of mixed computation and Futamura projections, Guntis noted that they used only program specialization, while ignoring the possibility of obtaining intermediate data. It would be interesting to look at the projections in more general interpretation of mixed computation.

For the sake of simplicity let mix computation be represented by two processors: specializer *spec* generating residual program p_x, and partial evaluator *peval* generating intermediate data x_p:

$$spec(p, x) = p_x$$

$$peval(p, x) = x_p$$

such that

$$p(x, y) = p_x(x_p, y)$$

for every y. On the one hand, if *peval* is trivial and x_p is always empty, these relations degenerate into ordinary mixed computation equation. On the other hand, if the result of available computation is expressed as intermediate data, we can consider the task of mixed computation to be converting the data x into another – in some sense, more effective – representation of x_p.

Substituting in the above equations p with *int*, x with program p, and y with the data d of program p, we have

$$spec(int, p) = int_p$$

$$peval(int, p) = p_{int}$$

such that

$$int(p, d) = int_p(p_{int}, d)$$

Besides converting a program into some intermediate representation, mixed computation also generates an interpreter of this representation. In the case of degeneration, generation of the interpreter int_p does not depend on program p at all, but only on the binding-time analysis result. The difference between such an interpreter and the initial one is the following: Whenever the initial interpreter performs computations over the initial data, the residual interpreter extracts the results of these computations from data structure x_p. The residual interpreter has some auxiliary variables pointing to the current data element of x_p. The algorithm we developed was the mirror image of polyvariant specialization, where the specialization process generated the next residual program fragment; the partial evaluator generated intermediate data element by putting there the values of this fragment's available expressions.

If available, we could split data into two sufficiently independent parts. There appears the possibility of a mixed strategy when we express one part of available computations in residual program and the other in intermediate data. The resulting residual program can be in turn specialized against the intermediate data. This strategy when applied to compilation problems corresponds to sequential *compilation phases*.

The main advantage of partial evaluation compared to specialization is that a more compact result occurs with the same volume of available computation because we no longer need to "decorate" computed values with the initial program fragments. On the other hand, we pay for this with the residual interpretation expenses [BarzBul88]. We did not bring the idea of polyvariant partial evaluator even to experimental implementation. Later Carolina Malmkjar from Copenhagen University did it.

Guntis Barzdyn wrote his Ph.D. thesis on the problems of inductive program synthesis. He successfully defended it after Ershov's death in 1988.

7   The M2Mix — Dmitry Kochetov

I became the scientific adviser to Dmitry Kochetov by recommendation from Igor V. Pottosin. The subject of his postgraduate research was a partial evaluator for a real-life programming language. Since Dmitry knew very little about mixed computation, he could not even imagine how difficult the task would be that he was going to attack. We chose Modula-2 as the source language. It was not only very popular at that time in the Institute of Informatics Systems. However, Modula-2 was also the language of the programming system that Igor Pottosin developed in the context of a contract with large industrial organization. That raised the chances of practical use of partial evaluator[6].

The M2Mix project required solutions of non-trivial theoretical and implementation problems. For example, since the source language had pointers and arrays, we needed a non-trivial binding time analysis and *alias analysis* in particular.

In order to avoid redundant code duplication, we developed an original *configuration analysis*. Its goal consisted of detecting for each program point the set of variables whose values influence specialization process. Even if the splitting of all program variables into static and dynamic does not change during specialization, the

---

[6] It should be noted here that partial evaluators were being already used in industrial projects. As an example, one can refer to the works of Samochadin [Samoch82] from Leningrad Polytechnic Institute on specialization of low-level control programs, the works of Romanovsky [Rom95] from the Institute of Automation of SB AS on the optimization of computer graphics programs. But in our case the objective was much more ambitious – to make practically useful a general purpose partial evaluator, which, like a compiler, cannot "know" about a particular application domain, but about the source language only.

trivial solution that presumes storing and comparing the completely active and reached memory makes such a partial evaluator practically unusable with respect to both time and memory. We can optimize the specialization process by discarding the following variables from consideration:

- o variables that do not change in the considered fragment of program. The internal representation of program in the interpreter specialization is an example of such variable;
- o variables that can be evaluated based on the values of other essential variables;
- o variables not used in the considered fragment, and dead variables in particular[7].

Similarly, it would be very unpractical to compare memory states at each source program point. In order to avoid non-termination it would be sufficient to trace only the set of so-called control points that cuts all program loops that modify static memory. This solution is evidently not optional. The problem is that a small number of control points may lead to residual code duplication, while large number of control points would lead to computational overhead.

Unlike most of the previous projects, we implemented M2Mix as generating extension processor. An important motivation for this decision was efficiency. However, the main reason was more fundamental; such implementation provided an easy way to guarantee that specialization performed exactly the same operations as the ordinary execution. Since we had no other way to execute a program but to translate it by the compiler at hand, we did both specialization and residual program execution in the same way[8].

One of specific features of Modula-2 language is the complete absence of unstructured *goto* control transfer. Being very positive for formal definition of analysis methods, it became a problem when it came to generation of residual program. To overcome that, we used a well-known trick of modeling labels and jumps by a global loop with nested switch.

Unexpectedly, it turned out that despite the fact that the residual program had minimum of interpretation left, it still worked several times more slowly than the original one. Again, the problem was in the fact that program efficiency depends not only on the number of executed high-level language operations, but also on the compiler's ability to implement them efficiently. In that respect, we could better optimize the compact and clear program of the original interpreter than the residual program with non-local control transfers and intertwined data dependencies. *Post-optimization* focused on this problem. It improved the residual program via common optimizations as well as the special transformations, which take into account the fact that the partial evaluator generated the program.

We tested the M2Mix on the standard test bench, including the Fast Fourier Transformation and the universal scanner Lex. These works formed the basis for

---

[7] The specified conditions are not precise and serve for demonstration of the idea only. For example, a variable may be used in the given fragment, but should be considered essential, since it must be transitively transferred to the fragment where it is used.

[8] Once we spent a lot of time trying to understand why some numerical analysis program and its specialized version produce different results. Finally, it turned out that the problem was in the fact that compiler options that were set for compilation of generating extension differ from those that were set for compilation of residual program.

Dmitry Kochetov's Ph.D. thesis "Efficient specialization of Algol-like programs". Currently, he works for Microsoft Corporation.

8 Binding-Time Improvement

In a certain sense, Nikolay Nepeivoda's prophesy came true. We had one of the most powerful partial evaluators, but had no program on which we could apply it successfully because when a programmer develops a general-purpose program he or she rarely cares about whether it is suitable for specialization or not.

The tools that improve source program binding-time properties would be a good supplement for further enhancement of partial evaluator. Boris Ostrovsky introduced a considerable selection of such transformations. Yuri Bannov in his master thesis has significantly extended this list and in addition, he has formulated conditions for appropriateness of their use. Following Boris Ostrovsky, Yuri Bannov worked on application of mixed computation to parsers, but his freedom of maneuver was restricted to partial evaluator M2Mix and universal parser Yacc[9].

The work titled "Polyvariant binding time analysis for higher-order functions" by Vladimir Ya. Kurlyandchik also focused on binding-time improvement. In contrast to the traditional orientation for the Novisibirsk community toward imperative languages, here the subject of the research was functional programs. The general idea consisted in the following: in usual monovariant binding-time analysis, we declare a variable as dynamic even if there is only one execution when it becomes dynamic. The developed methods allow for transformation of the original program by copying both program and data, so that more computations become static.

The university declared that the master theses of Vladimir Kurlyandchik and Yuri Bannov were the best graduate student's works in the recent years. Now they both work as managers in software companies.

9 Postscriptum

In 1992, the Institute of Informatics Systems organized the mixed computation research group. Later, in 1997 they transformed it into the laboratory of mixed computation. The tough 1990s seriously affected the character of research because financial self-provision became the highest priority. Experimental works, whose results could not have immediate use in industrial projects, were either frozen or continued by students. At the same time, computing machinery rapidly changed. New programming paradigms were widely adopted; the computing power increased by several orders of magnitude and they perfected the methods of compilation. Advanced compilers commonly used many features that seemed to be specific to mixed computation such as polyvariancy, loop expansion, and procedure unfolding.

---

[9] In the course of this work an interesting experiment was carried out that evidently proved the concept's efficiency. The task was to obtain two specialized parser: one should be generated automatically from the universal one, while the other should be written manually from the scratch. Even without mentioning that we could not get rid of all errors in manually written parser, it was simply less efficient than the one generated automatically.

However, it would be wrong to call the experience in the area of specialization without merit. A joint ongoing project between the St. Petersburg company TERCOM and American company Relativity Technologies [Terkom01] is an example. The project focused at developing a means for re-engineering and modernizing legacy applications. One of the key problems there is the *extraction of business logic*. One of the approaches, known as domain-based slicing, is essentially a specialization to known values of program variables in known program points. Even if the method is not suitable for obtaining of object code from an interpreter, it gives a satisfactory solution for many non-trivial tasks that emerge from practice and could unlikely appear in the "academic" setting. For example,

- o more than one value or a range of values may be associated with a variable for specialization,
- o negative specialization – specification of the set of values to which static variable should *not* be equal,
- o specification of the values of static variables not in the beginning of specialized program, but at the point where the values enter the program, e.g. where they are read from the database,
- o specialization of programs consisting of many components.

References

[BarzBul88] Barzdin, G.J., Bulyonkov, M.A.: Mixed Computation as a Tool for Extracting Compilation Phases, Methods of Compilation and Program Construction, All-Union Conference, Novosibirsk, pp. 21–23 (1988) (in Russian)

[Bul84] Bulyonkov, M.A.: Polyvariant mixed computation for analyzer programs. Acta Informatica 21, Fasc. 5, 473–484 (1984)

[Bul93] Bulyonkov, M.A.: Polyvariant binding time analysis. In: Proc. of the AC Symposium on partial evaluation and Semantics Based Program Manipulation, Copenhagen, pp. 59–65 (1993)

[BulErsh86] Bulyonkov, M.A., Ershov, A.P.: How Do Ad-Hoc Compiler Constructs Appear in Universal Mixed Computation Processes? Applied Logics 116, 47–66 (1986) (in Russian)

[Ersh84] Ershov A.P.: Mixed Computation. V Mire Nauki (6), 28–42 (1984) (in Russian)

[Itkin88] Itkin, V.E.: An Algebra and Axiomatization System of Mixed Computation. Partial Evaluation and Mixed Computation, 209–224 (1988)

[Jones88] Jones, N.D.: Challenging Problems in Partial Evaluation and Mixed Computation, Partial Evaluation and Mixed Computation. In: Bjørner, D., Ershov, A., Jones, N. (eds.) IFIP World Congress Proceedings, pp. 1–14. Elsevier Science Publishers B.V., North-Holland, Amsterdam (1988)

[Ostr87] Ostrovsky, B.N.: Controlled mixed computation and its application to systematic generation of language-oriented parsers. Programmirovanie 2, 56–67 (1987)

[Rom95] Romanovsky, A.V.: Mixed Computation application in Computer Graphics Problems, Author's abstract of Ph.D. Thesis, Novosibirsk (1995) (in Russian)

[Samoch82] Samochadin, A.V.: Optimized for Structured Microcomputer Assembler Programs. Microprocessor Programming, 89–99 (1982) (in Russian)

[Terkom01] Terekhov, A.A., Terekhov, A.N.: Automated Program Re-engineering. SPb State University, St. Petersburg (2001) (in Russian)

The Zelenograd Center of Microelectronics

B.M. Malashevich[1] and D.B. Malashevich[2]

[1] JSC "Angstrem
mbm@angstrem.ru
[2] National Research University, Moscow Institute of Electronic Technology (MIET)
denis@malashevich.ru

Abstract. This article deals with appearance of microelectronics in the USSR and establishing of its innovation centre in Zelenograd, Moscow. Prerequisites for creation of the microelectronics, measures on the development of the scientific-research centre are considered and the acting persons presented. Structure of the centre is described as a complex research and development corporation with complete set of functions. Its enterprises are displayed as well as their specialization and their first achievements. The technical level of the microelectronics centre production released in the first years of work is evaluated as corresponding to the world's manufacturing level; however, some reasons for lagging behind world leaders, beginning in the following period, are also mentioned.

Keywords: Centre of microelectronics, Zelenograd, (miniature) radio-set "Micro", IC, "Tropa".

1 Introduction

During the last half of the last century, the technology of assembly of the radio-electronic equipment (REE) from discrete elements has settled the possibility. The world had come to the sharpest REE crisis and radical measures were required. In the USSR the electronic industry was an independent branch of the State Committee on electronic techniques (abbreviated GKET in Russian), then transformed the Ministry of electronic industry (MEP), under the ministry of A.I. Shokin.

2 Preconditions

By this time and in the USSR and abroad, preconditions were already ripened for the creation of semiconductors and hybrid integrated circuits (IC). Integrated technology had been industrially mastered for semiconductor transistors, thick-film, and thin-film ceramic printed circuit-boards. The question was only: which one would be the first will be lit up with a pleasant IC idea. The first that appeared were Jack Kilby from Texas Instruments (TI) and Robert Noyce from Fairchild Semiconductor in the USA. In 1958, they had made the first IC: J. Kilby on germanium and R. Noyce on silicon. Juri Osokin of the Riga Plant of Semiconductor Instrumentation was the third one to begin production and supplies of the semiconductors ICs "R12-2" in 1962 (Figure 1).

J. Impagliazzo and E. Proydakov (Eds.): SoRuCom 2006, IFIP AICT 357, pp. 152–163, 2011.

Simultaneously with them appeared a hybrid IC. Both American and our experts predicted the most intensive development of the hybrid ICs, however, claiming that semiconductor ICs would dominate the market only by 1980.

Fig. 1. The three first ICs: of Jack Kilby, of Robert Neuse, and of Juri Osokin (an IC crystal with a germanium plate fragment in the background)

Using germanium for ICs was not promising. It has quickly understood both at TI and at Pulsar that the direction was toward silicon. Between 1959 and 1960, they began to work at Pulsar on creation of planar technology of silicon devices. By the end of 1961 at Pulsar, they generated a department of microelectronics whose chief was B.V. Malin.

3 Trailblazers

In the USSR, they formed two groups of trailblazers in the Soviet microelectronics: at GKET and at NPO "Almaz" (then "KB-1"). Since first half of the 1950s the Almaz chief engineer F.V. Lukin had organized active works on microminiaturization REE on the then available element base. However, by the end of 1950s, it became clear, that more radical methods were needed. Here then F.V. Lukin also had charged A.A. Kolosov to one of the most active, competent and interested in the decision of this problem of specialists in the Almaz. They freely knew three foreign languages thoroughly to study approaches to microminiaturization on foreign and domestic sources.

The results of this work were generalized in 1960 in A.A. Kolosov's small monographic called *Questions of Microelectronics*, which became the textbook for many specialists. In this work, the author has perfectly proved the necessity and timeliness of the beginning of large-scale works on research of the problems connected with the creation of integrated circuits; he stated new principles of REE creation. In 1960, F.V. Lukin had charged A.A. Kolosov to create the very first laboratory in the USSR on microelectronics. The laboratory started active work involving numerous scientific research institutes and universities as counterparts. Ideas of microelectronics had started to expand in the country. So, the non-suspecting F.V. Lukin, had started to prepare for a theoretical reserve and a staff for the Center of microelectronics in Zelenograd, which would be established in three years. See Figure 2 for a partial view of the center.

During the same period, A.I. Shokin's assembled a group of specialists at Pulsar and GKET to seek ways of generating output from the REE crisis. He had already concluded about the necessity of the creation and development of a new branch sector – microelectronics. The branch sector, (i.e. systems of scientific research institutes (abbreviated NII in Russian), would design offices (KB), and develop skilled and serial factories distributed all over the country; they would solve all special problems on the creation and duplication of products of

Fig. 2. Two of three buildings, from its beginning; Soviet Center of Microelectronics in Zelenograd

microelectronics. In 1959 he directed specialists to the USA to train and study planar silicon technology. When A.A. Kolosov had addressed K.I. Martjushchov's deputy, A.I. Shokin, with the results of the works and the offer on microelectronics, he received full understanding. At once, he had estimated the initiative and had suggested to organize a conference to gather the necessary heads. Such conference took place in the end 1961 in Leningrad under K.I. Martjushov's presidency. With the basic report A.A. Kolosov, with the supporting report on systems of memory, F.G. Staros, director SKB-2 in Leningrad, acted. Then A.A. Kolosov and K.I. Martjushov invited A.I. Shokin to discuss the problem, they came to the conclusion about the necessity of the creation of the uniform Center of microelectronics (CM).

The CM idea consisted in the formation of an innovative center for microelectronics. It would be locally placed functionally with a full complex of scientific research institutes with the experimental plants, solving all specific problems of creation and application of IC. For CM special materials, it should develop the technological, control, and measuring equipment for direct IC, based on REE. All this should be fulfilled on experimental plants and be transferred for mass duplicating to serial factories. In the USSR, they were already able to create the research-and-production centers and A.I. Shokin had the relevant experience.

4 The Decree

CM allocation had to be small and independent; the location city should be close to Moscow and it should create a scientific research institute and experimental plants. They found the place for such a city. In 1958, the Krjukovo railway station near Moscow was ideal for light industry construction of the "Sputnik" settlement (it has received the name «Zelenograd» in 1963).

By this time, there were obvious disproportions: provision for mass habitation was under construction and practically nothing had been made on the industrial building. However, expansion of works on creation CM needed the decree of the Central Committee of the CPSU and SM of the USSR; its output needed the consent of the first secretary of the Central Committee of the CPSU and chairman of SM of the USSR, N.S. Khrushchev. A.I. Shokin had begun preparations. The main base for preparation for the Decree the CM creation and all accompanying documents, posters, and exhibits became Pulsar. Supervising the preparation was V.N. Malin (chief of the general department of the central committee of the CPSU), I.D. Serbin (chief of the defensive department of the central committee of the CPSU), and L.V. Smirnov (chairman of the military-industrial commission).

In the beginning of 1962, A.I. Shokin had achieved N.S. Krushchev's consent to carry out a small exhibition with a report during a break of session of presidium of central committee of the CPSU. Hence, N.S. Khrushchev had already apprehended the idea and straight off did not reject it. Action took place, and N.S. Khrushchev had agreed on the further consideration of the offer. It has not simply agreed, but, similarly, had allocated for itself the problem of microelectronics and it was important for the nation.

Soon, in March of 1962, at the annual viewing of the architectural projects in Red hall of Mossoviet, they reported about serious disproportions in building of "Sputnik", N.S. Khrushchev said:

"It is necessary to discuss about microelectronics".

Apparently, he had discussed the issue with A.I. Shokin and F.G. Staros soon arrived at "Sputnik" for reconnaissance. In parallel with the preparation of the decree, work on the creation of a technology hybrid (in the SKB-2) and planar (in the Pulsar) integrated circuits were systematically developed.

For the final decision the situation in which would refer N.S. Khrushchev, microelectronics and demonstration of its advantages on clear was necessary to it an example. A.I. Shokin has created such situation. On 4 May of 1962 in Leningrad, the meeting with N.S. Khrushchev's participation on problems of ship construction was planned, and one of the major problems was onboard electronics. A.I. Shokin has applied all organizational experience and manager art, has involved old connection, and overlapping of the necessary events at last took place in time and space. Having arranged Khruschev's visit in SKB-2 by Staros before meeting, Alexander Ivanovich successfully used three trumps which in the given situation were directed at F.G. Staros: it KB was in Leningrad; at it was than to surprise; N.S. Khruschev was a bit attracted to F.G. Staros's since he accepted some participation in arrangement of its Russian destiny.

By the end of the 1990s in the American Russian-speaking magazine called *The Problem of Eastern Europe* appeared an article by M. Kuchmet titled "Participation of Americans in the Soviet microelectronics". The article affirmed that the Soviet microelectronics industry and its Center in Zelenograd's was created exclusively owing to the initiative, diligence and intelligence of two American engineers Alfred Sarant (in the USSR known as Phillip Georgievich Staros) and Joel Barr (known as Joseph Veniaminovich Berg). By 1950, they had emigrated from the USA to Czechoslovakia, and by 1955, they moved to the USSR where they headed a small

KB in Leningrad, where they were subordinated to the GKET. This article contradicted a reality that had started numerous publications and that had distorted the historical representation of CM's creation.

Because of F.G. Starosa's roles (I.V. Berg had no independent value), we shall dwell a bit more on this story. In SKB-2 they prepared F.G. Staros's then model samples of a control computer "UM-NH" and a tiny radio receiver. They were constructed based on tiny and non-packaged elements; the senior chiefs were amazed by the small sizes. The visit was well organized. Almost month of vigorous spadework proceeded. One day prior to A.I. Shokin's visit they led a rehearsal with Staros, which should be explanatory. The visit is well daring: "UM-NH" and a radio receiver had made a necessary impression upon N.S. Khruschev. At the same place, A.I. Shokin has reported the project of the Decree on creation CM in "Sputnik" as one approved in whole by Khruschev. After intensive coordination, on 8 August of 1962, they signed the Decree of the Central Committee of the CPSU and SM of the USSR. See Figure 3.

Fig. 3. Initiators creation Soviets microelectronics: Alexander Ivanovich Shokin, Feodor Viktorovich Lukin, Andrey Alexandrovich Kolosov, Boris Vladimirovich Malin, Fillip Georgievich Staros

As it is usual in similar cases, it was the conceptual Decree, the first in a turn that would follow behind. In it, they legalized that CM was to be in "Sputnik" and that henceforth, the problem of the creation and development of a homemade microelectronics plant has found the character of a national problem.

The general provisions of the concept of construction CM have been certain:

- Complex character CM with the organization of all cores of necessary scientific research institutes and experimental plants for designing and manufacturing IC is certain,
- CM the status of parent organization in the country on microelectronics with problems is given:
- Maintenance of designing and pilot production IC on a world technological level in interests of defense of the country and a national economy;
- Maintenance of a perspective scientific reserve;
- Development of principles of designing of the radio-electronic equipment and the computer on the basis of microelectronics, the organization of their manufacture, transfer of this experience to the corresponding organizations of the country;

- Unification IC, conditions of their application in the equipment at the enterprises of the country;
- A professional training, including specialists of the top skills.
- Local accommodation CM in "Sputnik", where CM becomes town organize system is certain.

The Decree made certain that an initial variant of structure of the CM enterprises. It had five new scientific research institutes with three experimental plants. They included the scientific research institute of theoretical bases of microelectronics, the scientific research institute of microcircuitry, the scientific research institute of technology of microelectronics, the scientific research institute of mechanical engineering, and the scientific research institute of special materials. Additionally, they were given corresponding tasks on their creation.

It is important to note once again that the creation of CM was not an isolated action; it was a part of larger program of creation new branch sector – microelectronics and A.I. Shokin was its initiator and its organizer of realization. In various regions of the country (Moscow, Leningrad, Kiev, Minsk, Voronezh, Riga, Vilnius, Novosibirsk, Baku and other places) saw the beginning of alterations of available GKET enterprise or the creation of new scientific research institutes with experimental plants. In addition, serial factories with KB emerged for design and mass production the ICs, special materials, and technological and control-measuring equipment. Thus, CM was only a part of a huge iceberg – a main peak, but only a part of it.

It is necessary to consider as special conditions of creation and development of domestic microelectronics. The electronic industry of the countries of the Europe, the USA, Japan what remained rigid was a competition between firms as were the participants of a wide international cooperation. Our microelectronics had been completely excluded from it. The USA had created the special international seventeen-country "Coordinating Committee of East-West Trade Policy" (COCOM) that supervised all scientific, technical, trade, and economic mutual relations from the USSR. The COCOM had developed a position paper of 250 pages across Soviet progress. It was impossible to sell not only high technologies and the products belonged to area of any high technology, microelectronics, and computers, but also other technological devices such as measuring equipment, materials, and precision machines. Consequently, in the Soviet electronic industry, it was necessary to do make everything ourselves. Certainly, special services partially managed to punch a wall on COCOM surrounding us and in a roundabout way to extract some products, documentation, materials, and equipment. However, all that was extracted was in scanty quantities and only to look and feel. It was necessary to design all this and to duplicate it in sufficient volume. Sometimes, the received samples were copied, but an exact copy to make it was impossible because of differences in materials, technologies, and equipment. Sometimes they did functional analogues; sometimes they completely did their own development. Nevertheless, they always developed and duplicated everything themselves.

5 The Center of Microelectronics

Soon after the release of the Decree, A.I. Shokin's command had started the creation of the CM (later the Centre of science – NC). The CM Decree had been given the right to employ specialists from any part of the USSR. Professionals and a scientific reserve in microelectronics, owing to the preliminary actions of A.I. Shokin, F.V. Lukin, and A.A. Kolosov who were already available in the country at the moment of signing of the Decree.

The formation of scientific research institute with experimental plants had begun. In 1962 began the scientific research institute of microdevices (NII MP) with a factory "Component" and Scientific research institute of precise machine building (NII TM) with "Elion". In 1963 began the scientific research institute of precise technologies (NII TT) with "Angstrem" and the scientific research institute of materiology (NII MV) with "Elma". In 1964 began the scientific research institute of molecular electronics (NII ME) with "Micron" and the scientific research institute of physical problems (NII FP). In 1965 began the Moscow institute of electronic technics (MIET) with "Proton" (in 1972). In 1968 began the central bureau on application of integrated microcircuits (CBPIMS). In 1969 began the specialized computer center (SVC) with "Logic" (in 1975). By the beginning of 1971 at NC in Zelenograd 12.8 thousand people were working. In 1976 based on the NC the country created the "Centre of science" with thirty-nine enterprises in different cities of the country that employed almost eighty thousand people.

The uniform organization at CM was not repeated; its enterprises submitted to the fourth central board of the GKET. The first had been organized by NII MP and NII TM. The appointed directors were I.N. Bukreev and E.H. Ivanov typed experts and organized work of scientific research institute while on the time premises. Workers of the Central board and GKET helped them. F.G. Staros also helped in part. Here is the recollection of I.N. Bukreev:

"Staros actively helped me. Specialists of NII MT trained at it in Leningrad. Besides in 1963 it had transferred us four designed it KB vacuum disposition system for mark thin-film (the first in the country). We at once began to master technology, and owing to it to 1964, there were the first microelectronic products. And if waited, while will construct our institute of mechanical engineering, we would lose two to three years".

Fig. 4. "Cub-2" - The module RON in capacity of 16 19-digit words, "Cub-2" in capacity of 128 19-digit words, Between them a coin 1 copeck 2003) The size of the module 32x34x4,125 mm. The size of the block: 128x19 bats – 32x34x42 mm 256x19 bats – 32x34x42 mm Such a Cub flied around of the Moon

F.G. Staros with CM shared both the ideas and a reserve. In NII MP, the idea of the micro-receiver was already based on microelectronic technology and it had been realized anew. Angstrem issued products designed in Leningrad such as the block of memory on ferrite plates with many openings of a "Cub-2", shown in Figure 4. In NII TT, the idea was more technological; a reliable variant of "Cub-3" was creatively advanced and designed. Probably, there were also other examples. F.G. Staros had really brought a certain contribution to the preparation of the creation of NC. He was one of the active members of the large command and this command stood at the forefront, though not the first.

On 29 January of 1963, the vice-president of the GKET approached F.V. Lukin and on 8 February of 1963, he became director of CM. Its deputy for science had appointed F.G. Staros, still remaining the chief at the Leningrad KB. However, F.G. Staros had ambitions and plans for the post of director of CM. Not having received his expected appointment, he had taken offence and had actually withdrawn from performance of functions of the deputy director. This caused harm to that business on which the organization had placed many forces. I.N. Bukreev tells about it in the following:

> *"The joke is recollected: «Have decided to collect those who on the first Saturday carrying a log from V.I. Lenin. Five thousand persons had responded nearby. Actually, there were only five. The same occurs now: many, "close known" Staros's, tell pull the long bow about its invaluable contribution to becoming Zelenograd microelectronics, initiating unhealthy interest to the person of this man. ... Itself Staros in Zelenograd never worked. I to it equipped a cabinet in "boarding school" (in 1-st microdistrict). But there was it here only 3-4 times at some o'clock. ... Actually, there was Staros in Zelenograd only till the summer a 1963. During this period from the first directors to it I contacted only. ... Lukin becomes director of a under construction Center of microelectronics and Staros's appoint it deputy on a science. ... Certainly, it wished to supervise over the Center. Well, does not happen. After purpose Lukin it more here did not appear."*

Director of NII TT of V.S. Sergeev is even more categorical:

> *"About Staros and Berg. I revolted when they named "father" of microelectronics and Zelenograd. From the point of view of techniques, their influence was zero. By the way, in America Staros in microelectronics was not engaged."*

These are direct certificates of two direct participants of theevents, the first directors of the first undertakings. One should not forget that Alfred Sarrant (*in the USSR - F.G. Staros*) stopped working as an electrical engineer in 1946, i.e. a year beforecreation of the first transistor in 1947 (the first informationabout it appeared in 1948), and in 1950 (before creation of the firstintegrated circuit in 1958) left the USA emigrating to Europe.

The heads created CM present on work just not twenty-four hours straight. Their drivers, who are not maintaining such mode of work that varied with frightening frequency. In reality, F.G. Staros's busy season happened in Zelenograd where he ignored weekly conferences. That is, in the fray of heavy daily work to creation the electronic industry, which demanded not only knowledge but also selflessness, he practically never participated. As a result, in 1964 there is an order on removing F.G. Staros's from the deputy director position at CM. Figure 5 shows the founders of the center.

Fig. 5. Founders of the Center of microelectronics in Zelenograd at an input in NII TT and a factory «Angstrem». From left to right: L.S. Garba (director of «Elma» at NII MV), B.V. Tarabrin (director of CB PIMS), K.A. Valiev (director of NII ME), A.J. Malinin (director of NII MV), V.F. Lukin (director НЦ), D.I. Juditsky (director SVC), A.K. Katman (chief engineer of NII TT), V.V. Savin (director of NII TM), G.V. Bechin (director of «Angstrem» at NII TT), V.S. Sergeev (director of NII TT).

6 The First Results

Soon, CM had started the creation of essential new production. In May of 1963 at NII TM they designed the first samples of vacuum deposition equipment. In the second half of 1963 at NII MP, the first results on thin-film technology were already developed and they designed the radio receiver «Micro». Its first model was the direct strengthening receiver and the second was the superhetrodyne. In 1964 when I.N. Bukreev brought this receiver to the USA at the congress of radio engineers, it made a world sensation! Photos and statements appeared in newspapers: How could the USSR overtake us? The «Micro» was sold for currency in France and England. Khruschev took them with himself abroad as souvenirs and gave them to G. Nasser and Queen Elizabeth.

The radio receiver «Micro» (Figure 6) used thin-film technology; it was first time in the country that they used batch production of microelectronics. In the second half of 1963 in NII MP the receiver was designed and its batch production began in 1964 at Angstrem's, where eighty thousand pieces were transferred to the MRP serial factory in Minsk. Up to the middle of the 1970s, one could buy this microreceiver in shops of the USSR and France.

M 1:1   M 2:1

Fig. 6. The product of microelectronics first in the USSR – radio receiver «Micro» It is produced on the basis of thin-film hybrid technology, a thin-film payment on a photo on the right with double increase. The size of the receiver 43x30x7,5 mm (without acting controls). The broadcast through phone for a hearing aid, inserted in an ear (the third at the left) was listened.

By 1964, NII TT started developing a series of thick-film HIC "Tropa"; its designer was A.K. Katman. Technical materials and literature in this area did not exist; there were only photos of the microcircuits that were issued by the firm IBM. All work began with nothing. They designed everything: the circuitry, the constructions, materials, technologies, and equipment.

By 1965 in Zelenograd, the Micron had begun to release the first semiconductor IC "Irtysh", designed by A.P. Golubev in NII ME based on the planar technology created in the Pulsar and put on the Micron. See Figures 7 and 8.

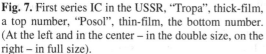

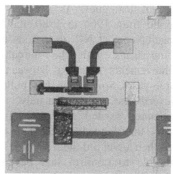

Fig. 7. First series IC in the USSR, "Tropa", thick-film, a top number, "Posol", thin-film, the bottom number. (At the left and in the center – in the double size, on the right – in full size).

Fig. 8. The first in CM semiconductor IC "Irtysh". Photo of topology of a crystal.

In 1966, Elma already issued fifteen kinds of the special materials designed in NII MV, and Elion produced twenty types of the technological devices in NII TM technological, control, and measuring equipment. In 1969, Angstrem and Micron already issued more than two hundred types ICs, and by 1975 in NC, they designed 1020 types ICs. They transferred all this to serial factories. This was only the beginning.

The results of many years of NC work did not look bad at a level of world microelectronics. Its first product – radio receiver «Micro», already had no equal the world. The first hybrid IC corresponded to a world level. By the way, the first-ever IC, flown the Moon (in 1969) and returned to the Earth, were Angstrem's "Tropas". In semiconductor IC, the USSR noticeably lagged behind, but would soon catch up with world leaders. In the 1970s, the most successful semiconductor company in the world was Intel. Intel issued the dynamic RAM in a capacity of 4K bits in 1974 and Angstrem issued the same in 1975. Accordingly, it issued 16K bits in 1977 and in the beginning of 1978; in 1979 both firms have issued on the market 64K-bit modules practically simultaneously. The similar situation existed at NII ME. In the beginning of the 1970s, the director of NII ME, K.A.Valiev, went to the USA to Motorola to show them the IC series 500 (analogous to the MC10000). Having investigated samples, experts of firm ascertained, that at identity of topology, IC series 500 had a higher speed than their designs and they were compelled to ascertain that the USSR technology was better. There are a number of other similar examples. See Figure 9.

The culmination of this competition began in 1979 when at NII TT they designed a single-crystal 16-digit computer, K1801VE1, with an "Electronics NC" architecture (in present terminology – the microcontroller). According to conclusion of the State Commission inspecting the project at that time, a computer of such level (with a 16-bit arithmetic-logical unit) did not have foreign analogues. As a whole during the period from 1964 to 1980, the backlog of development at NC on various types of IC comparisons with the maximum world achievements changed between zero and three years. Sometimes they pulled forward. Nevertheless, approximately the same dynamics existed and at leading foreign firms, they lagged behind their competitors and then advanced them a little. That is, it is possible to confirm that development of microelectronics at the NC in Zelenograd as a whole corresponded to a world level during those years.

However, regarding the volume of production of integrated circuits, the branch as a whole lagged considerably behind the

Fig. 9. IC "Tropa" first-ever send in 1969 in a free space, have flown about the Moon and have returned to the Earth. Fragment of a board of computer "Argon". Part IC is dismantled for research of results of long influence on them of a free space.

foreign level; the means did not suffice for the development of capacities of serial factories (and they in microelectronics were very dear) in the country. As a result, the loading on the experimental NC plants had sharply increased, specifically by a batch production of integrated circuits. This was the beginning of a fatal decline; it affected the further prospects of developing microelectronics. Opportunities to improve experimental plants for new materials, processes, technological routes, equipment, and products became sharply limited. Because of this and a number of other reasons, by approximately 1980 the progressing backlog of the Soviet microelectronics compared to a foreign level had begun. The rest is history.

Modular: The Super Computer

B.M. Malashevich[1] and D.B. Malashevich[2]

[1] JSC Angstrem, Moscow
mbm@angstrem.ru
[2] National Research University, Moscow Institute of Electronic Technology (MIET)
denis@malashevich.ru

Abstract. This article is deals with the history modular arithmetic, based on the rest classes calculus system «Restklassenarithmetik» (RCCS), or Modular Arithmetic, regarding its creation and development in the USSR. Characteristics of the USSR first modular super-computers: "T-340A", "K-340A", "Diamond (Алмаз)" and "5E53" and their development history are briefly considered. A brief analysis of the weakening of interest in the modular arithmetic in the USSR and its subsequent resumption of the Russian Federation is given. It also provides information on holding the 2006 special commemorative Scientific International Conference "50 Years of Modular Arithmetic" in the town of Zelenograd, Moscow.

Keywords: Modular Arithmetic, RCCS, K-340A, 5E53, Yuditski, Akushskiy.

1 Modular Arithmetic

In 1955 in Prague was published a collection of proceedings "Stroje Na Zpracovani Informaci", vol. 3, Nakl. CSAV that contains two articles: one by M. Valach and another by M. Valach and A. Svoboda. In these articles the idea to use operations on

$$M = p_1 p_2 \cdots p_n$$

residue ring, where p_1, p_2, ..., p_n is in pairs mutual-simple numbers, instead of the $M = 2^n$ residue ring operations on computer numbers initially advanced. The known Chinese remainder theorem, which before was treated as the structural theorem of abstract algebra, guaranteed the specified parallelism in calculations above integers provided that the result of ring operations belongs to the range of integers defined by product of modules $p_1 p_2 \cdots p_n$. As a result, the new position-independent notation formed has received the name «residue number system» (RNS). There was a new scientific direction in modular arithmetic based on RNS.

Soon the idea became known in the USSR. Compared with the sketchy information from different sources, it is possible to reconstruct this history as follows. Approximately

Fig. 1. Feodor Viktorovich Lukin

J. Impagliazzo and E. Proydakov (Eds.): SoRuCom 2006, IFIP AICT 357, pp. 164–173, 2011.
© IFIP International Federation for Information Processing 2011

1957 regarding the "Almaz" (then "KB-1") the information on works in the USA on modular arithmetic arrived. The chief engineer of the Almaz, F.V. Lukin (Figure 1.), had personal experience in the design of computing devices and, especially, of their application in the largest systems. He appreciated perspectives of this direction. Feodor Viktorovich was the main organizer and the patron of works on the development of modular arithmetic. His diligence received a robust and successful development in the country. The beginning of its demise coincides with his departure from life.

Notwithstanding, the Almaz was not an engagement in the design of a computer and F.V. Lukin had sent the inquiry to the scientific research institute of electronic machines (in Russian - NII EM). In 1953, he was their chairman of the state commission on acceptance of computer "Strela", whose first appearance had been established in the Almaz. The inquiry had interested the mathematician Israel Akushsky and his chief leading the design of computer Davlet Juditsky, to become the founders of modular arithmetic in the USSR. One of leading theorists in the RNS method and active participants of its practical application (Dr.Sci.Tech.), professor and academician Academy of Sciences of Kazakhstan, Viljan Amerbaev recollects:

> "Israel Akushsky told to me, that first information about RNS he has received from F.V. Lukin in the form of the inquiry on works in the USA. According to Israel Jakovlevich, Feodor Viktorovich considered RNS as very perspective direction of development of computer facilities".

After this, the articles by A. Svoboda and M. Valach had arrived to I.J. Akushsky for preparation of its abstract. The initial information received (rather brief and superficial) has started the scientific research of I.J. Akushsky and D.I. Juditsky. The first attempt in the country to comprehend principles of construction modular computer was undertaken between 1957 and 1958 in NII EM by J.J. Bazilevsky, J.A. Shrejder, I.J. Akushsky, and D.I. Juditsky. However, it had not received uniform understanding and not all participants had taken a liking toward the essence of RNS.

2 Super-Computer "T-340A" and "K-340A"

In 1960, F.V. Lukin was recently nominated the director of the scientific research institute NII DAR (then the NII-37). He invited D.I. Juditsky and I.J. Akushsky to design the computer. D.I. Juditsky became the chief of the department and I.J. Akushsky became the chief of the laboratory in this department.

Between 1960 and 1963 in this department, the first modular super-computer in the country (the T-340A) was designed as an experimental model of the complicated system. (Here, a super-computer is understood as a computer with record-breaking high characteristics at that time). The theory and practice of variant modular arithmetic principles of construction of the computer was designed on this basis by I.J. Akushsky, D.I. Juditsky, and E.S. Andrianov. This computer had really worked for many years in the experimental system. The received

Fig. 2. Israel Jakovlevich Akushsky

results have been used in designing the K-340A computer, which has been mastered in a batch production and became the base for all systems designed in those years within NII DAR. In these computers, they released the principles of independent commands and data memory channels.

Operative memory was executed in the form of 16 blocks each with 1K-word capacity. Each block had two ports for input-output of information with subscribers (with an opportunity of a parallel exchange with any number of blocks) and with the processor. For speed increase, they realized program stratification of operative memory with alternation of the reference of the processor to blocks. In addition to the multiport, they applied buffer memory to two-operational commands. (Each command was carried out on two operations, each of which in other computers of that time was executed in the form of a separate command.) These features of memory system construction have provided high efficiency of the computer; delays were practically non-existent regarding the reference to memory of great volume (a scourge of the computer of those years). The speed of the T-340A and K-340A computers reached 1.2 million doubled, or 2.4 million single operations per second (OPS). Typical speed of the computer in those days was measured by tens or hundreds thousands OPS. By two factories it has been able to produce more than fifty K-340A computers, some of which in the structure of systems are still in operation until now - in 2006 it is forty years!!

Table 1. Specifications for the "T-340A" and "K-340A" Computers

The main designer:	T-340A - D.I. Juditsky, K-340A - D.I. Juditsky, after L.V. Vasiljev.
Development: NII DAR:	T-340A - 1960 ... 1963 K-340A - 1963 ... 1966
Manufacturers:	An experimental plant at NII DAR and the Sverdlovsk factory of radio equipment, per 1966-1973 it is let produce more than 50 complete sets.
Word length of data and commands:	45 bit
Notation:	Residue number system (RNS)
RNS - the bases and order of a word score by them:	
The bases:	2; 5; 23; 63; 17; 19; 29; 13; 31; 61.
Order of a word:	1; 2-4; 5-9; 10-15; 16-20; 21-25; 26-30; 31-34; 35-39; 40-45.
Performance:	1.2 million two-operational OPS (in the standard calculation, up to 2.4 million OPS)
Detection of an error in a word:	At performance of operations in the arithmetic device
Multiport buffer memory:	16 x 45 bit
RAM:	16K 45-digit words (720K bit)
ROM of commands:	16K 45-digit words (720K bit)
Element base:	Transistors, diodes, ferrite, etc.
Power consumption:	33 KW
The size of a rack cabinets:	600 x 700 x 1800 mm
Number of rack cabinets:	12

3 Super-Computer "Almaz"

In the beginning of 1963, F.V. Lukin was appointed director of the Center of microelectronics (CM), built in Zelenograd. One of its primary goals was:

> *"Design of construction principles of the radio-electronic equipment and the computer on the basis of microelectronics, the organization of their manufacture, transfer of this experience to the corresponding organizations of the country".*

To perform this task, Feodor Viktorovich invited his well-known collective of the T340A and K340A computer founders led by D.I. Juditsky and I.J. Akushsky. By this time the T-340A computer had been designed, constructed, and modified. The design of the project of serial computer K-340A, its manufacturing and debugging on an experimental plant of NII DAR have been completed after leaving the group of specialists in Zelenograd by remained collective of employees under direction of Leonid Viktorovich Vasiljev. In 1964, they had formed a department of perspective computers in the enterprise known nowadays as scientific research institute of physical problems (NII FP). D.I. Juditsky was nominated the chief engineer of NII FP.

Fig. 3. Davlet Islamovich Juditsky

By the end 1965, three organizations were given the competitive task to design outline sketches of the high-performance super-computer with a release date of 30 March 1967. The organizations were (i) the center of microelectronics (the ministry of electronic industry (MEP), designer F.V. Lukin), (ii) the institute of exact mechanics and computer facilities (the ministry of the radio industry (MRP), designer S.A. Lebedev), and (iii) the institute of electronic operating machines (the ministry of apparatus making, designer M.A. Kartsev)

The computer was to include the following characteristics: data word length of 45 bits, performance of 2.5 to 3.0 million algorithmic OPS, complex functions in one command (one algorithmic operation on tasks of the customer on the average corresponded to the four usual operations of the computer), work with words of variable length, and a memory size of 2^{17} 45-bit words (5.625M bits). That is, the usual understanding that required a computer with speed nearby 10 (9 to 12) one million OPS. Best known, by the end of 1966 computers possessed speed in 4 to 12 times smaller demanded the following.

Firm	Model	Speed of the computer, one million additions/second	Speed of elements, nanosecond
IBM	360/75	1,0	5
CDC	6600	2,5	10
Philco	2000/212	1,5	5
Burroughs	B 5500	0,3	20
Sparry Rand	1108	1,2	5

So in Zelenograd, the design of the outline sketch of super-computer "Almaz", designer D.I. Juditsky had begun.

All forces of CM were involved in creating the "Almaz" computer, shown if Figure 4. NII FP executed development of architecture and the processor of the computer; NII TM did the base construction, the power supply system, and input/output system; NII TT designed the integrated circuit (IC). In this aspect the project Almaz had conclusive advantage in comparison with S.A. Lebedev's and M.A. Kartsev's projects, since the newest element base was created here, in Zelenograd, and on this process it had great influence.

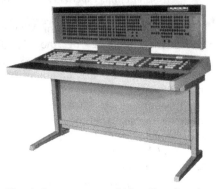

Alongside with the application of modular arithmetic, they found one more architectural way to obtain substantial growth of the general productivity of the computer. This was the decision widely applied in later systems of processing of signals by the introduction of a system of the processor of preliminary processing a signal. Soon, it became a new word in a science and technique. Into structure of the "Almaz" computer, they used three types of computing processors.

Fig. 4. Super-computer "Almaz" engineering control console

Table 2. Specifications for the "Almaz" Super-Computer

The outline sketch:	March I968
Main designer:	D.I. Juditsky,
	Supervisor of studies I.J. Akushsky
The developer:	Center of microelectronics, Zelenograd
Word length of data and commands:	45 bits
Performance:	7.5 million algorithmic OPS (in the standard calculation, up to 30 million OPS)
Notation:	Residue number system (RNS) with the additional basis
RNS - the bases and order of a word score by them:	
The bases:	2; 5; 23; 63; 17; 19; 29; 13; 31; 61.
Order of a word:	1; 2-4; 5-9; 10-15; 16-20; 21-25; 26-30; 31-34; 35-39; 40-45
Detection of double and correction of single errors:	Performance of operations in the arithmetic device, hardware majorization (2 from 3) in all other devices
Calculation value of special functions:	Used as an elementary command
Memory size:	128K 45-digit words (5,898M bits)
Fast buffer memory:	32 55-digit words
Element base:	IC series "Tropa"
Power consumption:	5 KW
The size of a rack cabinets:	550x800x1750 mm
Volume of the equipment:	11 rack cabinets, an engineering control console, external devices
The occupied area:	80,100 м^2

o A narrowly specialized non-programmable processor for preliminary processing the information acting in real time, named in the Almaz the Converter of the information (CI);

o A programmable modular processor which is carrying out the basic data processing;

o A programmable binary processor that carries out non-modular operations used for computer control procedures.

The information from a source comes to CI, passes preliminary processing in real rate of receipt that excludes necessity of its intermediate storage. The results of this processing (their volume repeatedly less initial) act on the modular processor. Calculations have shown, that offered CI has the productivity equivalent about 4.0 million algorithmic OPS and allows to save about 3 megabits of memory. The modular processor of the "Almaz" computer has a productivity of 3.5 million algorithmic OPS. As a result, effective productivity of Almaz makes 3.5 + 4.0 = 7.5 million algorithmic OPS; that is, up to two to three times above requirements. See Table 2. These settlement data were confirmed by results of modeling on a universal computer and experimentally on the sample of the "Almaz" computer.

4 Super-Computer "5E53"

The outline sketch has been designed and on March, 30th, 1967 is presented to the customer. Competition has won computer "Almaz". In May, 1967 CM has received the order for design of the high-performance super-computer "5E53" and a 5-machine complex on its basis with the organization of a batch production in Zagorsk city an electromechanical factory (ZEMZ) MRP. The main designer 5E53 had been nominated D.I.Juditsky. In October, 1969 the collective of designers of the computer has been allocated in the independent organization - Specialized computer center (SVC), director D.I.Juditsky, the deputy science on I.J.Akushsky.

While Almaz was designed, the customer has specified characteristics: the 40-bit computer with performance on tasks of the customer up to 10 million algorithmic ops (nearby 40 million usual ops), the RAM 7,0M bit, PROM 2,9M bit, external memory 3G bit, the equipment of data transmission on hundreds kilometers was required.

The architecture 5E53 had many progressive decisions:

o Division of commands on administrative and arithmetic. Arithmetic commands (including preliminary and basic processing of the information) were carried out on modular processors, administrative - on binary.

o 8-levels the conveyor organization.

o Hardware block realization of arithmetic's: the block of addition/subtraction, the block of multiplication, the block of management of addresses, etc.),

o Division of data RAM and commands PROM,

o Division of trunks of commands bus and data bus,

o Hardware stratification of memory on 8 blocks with an alternating addressing on blocks.

Special RAM based on the integrated carrier (cylindrical magnetic film) was designed for the 5E53. By the speed, dimensions, weight, power, and cost it was much more attractive than an adaptability to manufacture applied RAM on ferrite cores. See Figure 5.

Fig. 5. Fragment of the pilot sample super-computer's 5E53

One more of the main problems was construction of PROM for storage of programs and constants which on tasks of the customer vary not often. Simple enough and fast constant memory, but supposing change of the information therefore was required. For 5E53 it is developed PROM with replaceable induction cards.

The storage on an optical type was designed as external memory of a large capacity. It had much in common with the cores at that time external memory on magnetic 35 mm tapes (a similar design, a drive, electronics), but differed in a data carrier and methods of record/reading of the information; it used photo/light-emitting diodes through an optical fiber on a film. As a result, the capacity of external memory increased by two orders of magnitude and attained 3G bits. The model of the storage was produced and it worked in the structure of the pilot model of the 5E53.

Attained reliability of the 5E53 was provided with self-corrected properties RNS in the arithmetic device, full majorization (2 of 3) all other systems of the computer's technology of installation intercell and interblock connections by a method turning and other means.

In the beginning of 1971, the design of the documentation was completed. All necessary tests of cells and subassembly have been finished and the pilot model of the 5E53 was fabricated and tested. On 27 February of 1971, eight complete sets of the design documentation was produced. Preparation of manufacture had begun. However, in the beginning of 1972 when preparation of manufacture 5E53 already came to the end and there has been begun manufacturing of separate devices of the computer, work on creation of system for which it intended stopped. Simultaneously,

Table 3. Specifications for the "5E53" Super-Computer

The contract design:	February 1971
Main designer:	D.I. Juditsky
Leading developers:	V.M. Amerbaev, I.J. Akushsky V.M. Radunsky, L.G. Rykov, M.N. Belova, P.F. Silantev, J.N. Cherkasov, V.S. Butuzov, V.A. Merkulov, P.V. Nesterov, V.N. Shugin
The developer:	Specialized computer center, MEP, Zelenograd
The manufacturer of the pilot sample:	Specialized computer center (SVC)
Word length:	Data - 20 bits and 40 bits Commands - 72 bits
Notation:	Residue number system (RNS) with the additional basis
The bases:	17; 19; 26; 31; 23; 25; 27; 29
Order of a word:	1-5; 6-10; 11-15; 16-20; 21-25; 26-30; 31-35; 36-40
Clock frequency:	6.0 MHz
Performance:	10 million algorithmic operations a second on tasks of the customer (40 million OPS) 6.6 million OPS on one the modular processor
Format of algorithmic operation:	Average is four usual
Time of performance modular operations:	1 step is 166 nanoseconds
Number of processors:	8 (4 modular and 4 binary)
PROM command:	Capacity: general is 2.8M bits A case is 573K bits The block is 72K bits Time of a cycle 332 nanoseconds, Rate of sample - 166 nanoseconds, Number of blocks - 40, Number of rack cabinets - 5.
The RAM of data:	Capacity: general{common} 7.0M bits A case is 1.0M bits The block is 4096 x 64 = 256K bits Time of a cycle is 700 nanoseconds Rate of sample is 166 nanoseconds Number of blocks is 28 Number of rack cabinets is 7
Volume of the equipment of the computer:	Types of rack cabinets is 7 and the engineering control console Number of rack cabinets is 24
The size of a rack cabinets:	1800 x 800 x 600 mm
Element base:	IC series "Tropa", "Posol", "Krug".
Power consumption:	60 KW.
Average time of non-failure operation:	600 hours
The occupied area (with the bench and repair equipment):	120m^2

work on the 5E53 stopped. There was no other consumer or other manufacturer in MRP. Although there was no demand for the 5E53, it appeared that in MEP the tasks for it continued. Low integration ICs could be designed with manual design; the time for large ICs and powerful systems of the automated designing has not yet arrived.

5 Destiny of Modular Arithmetic

In the 1960s and 1970s with respect to the designers of super-computer A-340A, K-340A, Almaz, and 5E53 in NII DAR, SVC and in the enterprises cooperating with them, serious scientific researches in the field of modular arithmetic were made. There were many publications on this theme in the open press, including and in the form of monographs. They had aroused serious interest at domestic and foreign specialists.

The true reasons for stopping the 5E53 project practically nobody knew. In fact, having received wide publicity in circles of experts and appreciated by them as a failure of the project, there began an independent life. There was an almost insuperable barrier on the further ways of introducing RNS in domestic computer facilities. Further use of modular arithmetic in the USSR energized enthusiasts, basically, in the theoretical plan and they were engaged only in the arithmetic. Foreign specialists of such shock have not gone though and there modular arithmetic has received more consecutive development. Figure 6 shows designers of the popular modular super-computers.

Fig. 6. V.S. Kokotin, M.D. Kornev, M.N. Belova, L.G. Rikov, V.S. Khajkov. Designers of the modular super-computers T340A, K-340A, Almaz, and 5E53. (22 September 2004 in Zelenograd, for D.I. Juditsky's 75-year birthday).

Currently, activity is characterized by two moments. First, there appeared to be a sharp increase in demand for the decision of tasks with prevalence of modular operations such as in processing of signals, images, cryptography, and toponymy, in addition to a demand for high reliability and productivity. On these tasks, modular arithmetic is also effective. Secondly, change the principles of creation of integrated circuits; decomposition of this process on "front-end" stages (circuitry designing) and "back-end" (layout designing), development of technology of the IP-blocks, programmed logic integrated circuit (PLIC) and a line of other innovations in the microelectronics, new devices facilitating integrated construction. As a result, interest in modular arithmetic again increases. The volume of publications has considerably increased for this theme and many enterprises have begun researches in the field of modular arithmetic.

The year 2005 marks the fiftieth anniversary of RNS based on modular arithmetic. It is a good occasion for summarizing its development, an estimation of a modern condition and prospects of development. Therefore, a number of the enterprises from five countries (Russia, Kazakhstan, Ukraine, Belarus, and the USA) have made a decision on carrying out of anniversary special scientific international conference "50 years of modular arithmetic". The materials of the conference appear on the site of the Moscow Institute of Electronic Technology (Technical University) (MIET)[1] and the Virtual computer museum[2] also are published in the form of the collection of proceedings of "50 Years of Modular Arithmetic". The anniversary international scientific and technical conference took place within the program of the International scientific and technical conference "Electronics and computer science - 2005". The collection of proceedings appear in Open Society "Angstrem", MIET, 2006, page 775 with ISBN 5-7256-0409-8.

The proceedings presented on conference testify that for fifty years, modular arithmetic has developed within independent scientific schools with various directions and real applications.

---

[1] http://www.mocnit.miee.ru/oroks22W
[2] http://www.computer-museum.ru/ and
<http://www.computer-museum.ru/histussr/sokconf0.htm>

The Microprocessors, Mini- and Micro-computers with Architecture "Electronics NC" in Zelenograd

B.M. Malashevich

JSC Angstrem
mbm@angstrem.ru

Abstract. This article deals with the history of research, design and development of the mini-computer family, microprocessors and micro-computers with the so-called SC - architecture (from *Scientific Centre* in Russian writing — abbreviated from 'Nauchnyi Centr' (НЦ — Научный Центр) and the based on them systems, created by the Zelenograd Microelectronics Centre in the first half of the 1970s. The «Electronica NC» family ("Электроника НЦ") had a bus-modular structure, that enabled easy creation of various computation, control, and management systems. At the first development stage the family modules were based on integrated circuits with low and medium integration levels (ICs and MICs). At the second stage, those were BIC microprocessors, memory storages, and BICs based on basic matrix crystals (BMC - chips).

Keywords: «Electronica-NC» (*Электроника НЦ*), «Yuryusan'» (*Юрюзань*), high-frequency communication channels «Svyaz-1» (КВС *Связь-1*), microprocessor.

1 Mini-computers and Mini-systems

1.1 Electronics NC

In the beginning of 1973 D.I. Juditsky, director of the specialized computer center (SVC) in Zelenograd, has assembled a compact working group composed of D.I. Juditsky, M.M. Khokhlov, V.V. Smirnov, B.A. Mikhajlov and J.L. Zakharov to design the architecture of a mini-computer – a new direction of development in SVC. They analyzed the best foreign and domestic experiences, collected all perspective ideas, added their own ideas, and harmoniously synthesized these traits into a uniform architecture for construction of some compatible mini-computers and systems. Based on this work, it has received the name "Electronics NC" from the "Centre of science" (abbreviated NC in Russian) – the name of the center of microelectronics in Zelenograd, within the SVC. The basis of the work included a bus-modular structure, a micro-program controller, a version of architecture based on a program by means of PROM logic, a base kernel of system of commands with a reserve for applied expansions, the modular software, a test system of self-diagnostics, cross-system programming on universal BESM-6 and ES EVM computers, and a number of other progressive characteristics.

J. Impagliazzo and E. Proydakov (Eds.): SoRuCom 2006, IFIP AICT 357, pp. 174–186, 2011.
© IFIP International Federation for Information Processing 2011

1.2 Electronics NC-1

In the same year of 1973, they designed, made, and handed over to state commission a modular reconfiguration of the minicomputer "Electronics NC-1". Its main designer (MD) was D.I. Juditsky with other designers such as M.M. Khokhlov, V.V. Smirnov, B.A. Mikhajlov, J.L. Zakharov, V.S. Kokorin, A.M. Smagly, V.A. Merkulov, V.N. Shmigelsky, P.P. Silantjev, A.V. Bokarev, V.M. Trojanovsky, B.V. Shevkopljas, and F.I. Romanov. Figure 1 shows the NC-1 computer.

The NC-1 was a 16-bit control computer with a speed up to 0.7 million OPS. It used integrated RAM modules with a capacity of 128K byte on cylindrical magnetic film and a PROM of 7K byte on replaceable induction cards. The computer had functional and construction modular structures, allowing one to complete various systems. The interface for input-output provided a connection for peripheral devices of the cores then in the country of families of the computers such as ASVT and ES EVM. They developed other computers such as UVO, SUPVV, VSU, and UKPO (see below). They mastered a batch production of the NC-1 (1974-1989) by the Pskov factory of radio components. Later in Pskov had made variants of the computer of LSI and issued it under names "Electronics NC-2" and "Electronics 5337".

Fig. 1. Mini-computer "Electronics NC-1"

1.3 DSC "Jurjuzan"

By the end 1972, SVC had received an order from the ministry of civil aircraft (MGA) of the USSR to develop a data-switching center "Jurjuzan" (DSC). They would install a prototype at the airport of the Pulkovo in Leningrad. A subsequent batch production (MD D.I. Juditsky, then V.S. Butuzov with developers such as N.A. Smirnov, V.S. Sedov, V.S. Travnitsky, A.N. Lavrenov, and N.K. Ostapenko) would also occur. The Jurjuzan represented four-machines (four NC-1 computers) and a duplicated two-channel hardware-software complex. Each channel consisted of a computer for interaction with line channels, a computer for processing telegrams, and equipment for communication with line channels. The DSC provided processing 64 telegraph channels with automatic check and correction of telegrams. The set of NC-1 modules had filled up by a multiplexer for data transmission. See Figures 2, 3, and 4.

By November of 1976, the DSC had been developed, made at the "Logic" factory at SVC. They modified it and then entered it into pre-production operation at Pulkovo. However, in the middle 1976, they transformed the NC in Zelenograd into a larger NPO "centre of science" (NPO NC). Based upon the SVC and the NC management, it created a special design bureau "centre of science" (SKB NC), the headquarters plant at NPO NC. SKB NC did not engage in the design of production. Actually, in fact, they liquidated the SVC and transferred its designers to the Logic division of the scientific research institute of precise technologies (NII TT). This was the "Angstrem" factory at NII TT; D.I. Juditsky departed NC. The new management had categorically refused to duplicate the DSC. The DSC «Jurjuzan» in Pulkovo has worked almost twenty years until 1995.

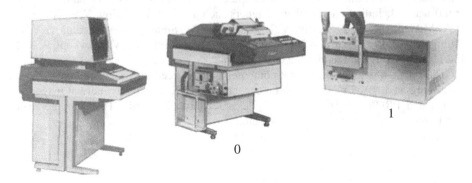

Fig. 2. The device of visual display (UVO). The symbolical display. Diagonal of the screen (43 cm); Size of the screen (220x200 mm); Symbols (up to 2048); Lines on the screen (32); Symbols in line (64); Size of a symbol (3.5x2.5 mm); Ensemble of symbols (128).

Fig. 3. The combined device of preparation and input-output of the information (SUPVV). SUPVV included Tape puncher (PL-150); Photo input reader punched tapes (FS-1501); Printing machine type «Consul-260»; Modes (independent and from the computer).

Fig. 4. The compact-cassette store on a magnetic tape (KNML). Type of the cartridge (NK-60); Information capacity cartridges (5 Mbit); Speed of an exchange (5680 bps); Length of a file of information (any); Hardware control of information; Interface (ES EVM).

1.4 CAC «Svjaz-1»

In August of 1974 under order LNPO «Krasnaja zarja», the SVC had begun to design the computing aids complex (CAC) «Svjaz-1» (MD D.I. Juditsky; designers include A.A. Popov, N.M. Vorobjov, V.A. Gluhman, A.P. Seleznev, M.D. Kornev, V.A. Merkulov, V.A. Savelichev, and A.I. Koekin). See Figure 5. They used the NC-1 and DSC hardware and program modules, but they also designed new modules. The Svjaz-1 had a universal purpose with a wide spectrum of variants of configurations (from 1 up to 30 processors) with differing computing resources. It provided maximal efficiency and survivability by means of computing process parallelism, a popular accessible field of memory, reconfigurable structure, and hardware duplication of computing process. They carried out the role of the central operating body in the CAC with a modular OS. Having finished the current task, each processor of system carries out tasks independently addressed in a table of tasks and received from turn the new task (including and a role of the main processor). Each module had some variants of ways to reference any other module that allowed one to use flexibly resources of system and provided its high survivability; a refusal of a part of modules led only to decline of productivity of system.

The CAC and it software have been developed, the project is accepted by the customer, the design and program documentation in second half 1976 are transferred to the «Krasnaja zarja» for a batch production. However, because of the liquidation

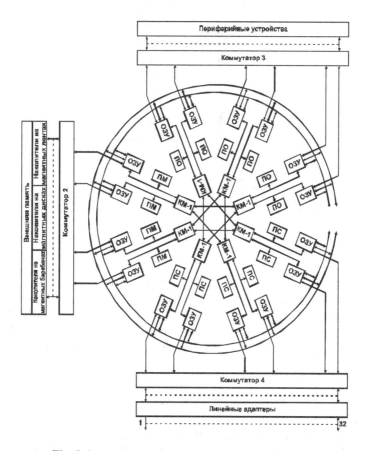

Fig. 5. Structure of 16-processor variant CAC "Svjaz-1"

SVC, NPO NC had refused continuation of work. The CAC with some completions without participation of the designers under the name "Svjaz-M" has been mastered by the «Krasnaja zarja» in a batch production, was issued for a number of years and there was base CAC for many communication systems which are designed and let out those years LNPO «Krasnaja zarja». See Figure 6.

Fig. 6. SUBK-SM – one of configurations CAC "Svjaz-M"

2 Microprocessors

The NC architecture used the development principles of microprocessors and microcomputers. The design came about by SVC specialties (architecture, circuitry) complete microprocessor sets (MPC), the LSI series (587, 588, 1801, 1802, and 1883), in a number of the micro-computer Electronics (NC-01, -02, -02M, -03T, -03Д, -03S, -04T, -04U, -05T, -8001, -8010, -8020), and in systems Electronics NC-31, YW-32, and "Tonus NC-01".

It began in 1973 when D.I. Juditsky was charged to organize a youth collective of V.L. Dshkhunjan's laboratory to borrow a study of approaches to construct microprocessors. Participants were V.L. Dshkhunjan, V.V. Telenkov, P.R. Mashevich, J.I. Borshchenko, V.R. Naumenkov, I.A. Burmistrov, S.S. Kovalenko, and A.R. Tizenberg. This collective, with the active help of leading SVC specialists, had designed original architecture of the sectioned MPC the LSI. Designed the LSI was carried out in close cooperation with semiconductor firms on circuitry to projects SVC developed topology and a manufacturing techniques the LSI. Thus, they created five MPC LSI sets on the basic for those times microelectronic technologies:

- КМОП (9-volt) – series K587, SVC, NII TT and Angstrem,
- КМОП (5-volt) – series K588, SVC, NII TT and NPO "Integral",
- TTL – series K1802, SVC, NII TT, NII ME and Micron,
- nМОП – series K1801 (its the first the LSI) in NII TT,
- nМОП – series K1883 (in GDR – U-83), SVC, NII TT and Robotron.

Figure 7 shows the topology of one such processor.

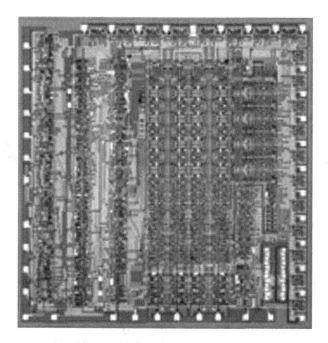

Fig. 7. Topology K587IK2

These series (except for single-chip K1801) represented the partitioned microprocessor complete sets with the same architecture of the open type, allowing to design on them various microcomputers and systems.

The first-born and an example of design of these complete sets the LSI is MPC series K587 for construction of the computers and systems with word length of data, multiple 4 bit which became a basis for the first the micro-computer in Zelenograd:

- K587IK2 – 4-digit section of the arithmetic device.
- K587IK3 – 8-digit section of an arithmetic dilator.
- K587IK1 – 8-digit section of information interchange.
- K587RP1 – section of memory for blocks of microprogram management.

Figure 8 shows the series.

K587IK2   K587IK1   K587IK3   K587RP1

Fig. 8. The first in USSR microprocessor complete set, a series K587. It was applied by consumers about twenty-five years.

3 Micro-computers and Micro-systems

3.1 Electronics NC-01

In 1974 it is designed and 2 samples produced of first samples LSI of series K587 one-board 16-bit micro-computer "Electronics NC-01" (Figure 9) with speed 250 000 ops, from the RAM 1K byte and with two parallel programmed input-output ports of data. (the MD was D.I. Juditsky with designers such as V.N. Lukashov, A.A. Popov, J.M. Petrov, and V.A. Merkulov.

Fig. 9. Electronics NC-01

3.2 Electronics NC-02, НЦ-02М

In 1975 we saw the design of the microcomputer "Electronics NC-02" (MD was J.M. Petrov with designers that include V.N. Lukashov, A.A. Popov, J.M. Petrov, and V.A. Merkulov). See Figure 10. It was a two-board 16-bit computer in the compact case with a mobile control panel. As a board of the processor, it became an updated NC-01. On the second board appeared the adapter of interfaces. Between 1976 and 1977, Logica and Angstrem produced more than forty NC-02 used in the technology equipment. In 1976, they began modernizing the computer. In the new case, there were empty slots for installation of additional on-board devices. In total, they produced sixty-three NC-02M computers, which they applied in a variety of technological equipment.

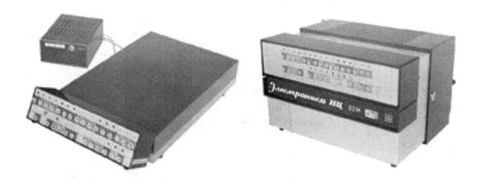

Fig. 10. Electronics NC-02 and NC-02M

The NC-01, NC-02, and NC-02M microcomputers actually were laboratory models for a working architecture, a design, a design technology, and for manufacturing microcomputers at a time when it was essentially a new kind of product.

The accumulated experience and study of foreign novelties have allowed the team to complete the perfection of the NC architecture and based on it, to design the architecture of three software and hardware compatible "Electronics NC" microcomputers with a consecutive increase in computing capacity; they are the

NC-03, NC-04, and NC-05. The architectural designers included D.I. Juditsky, N.M. Voobjov, M.D. Kornev, A.A. Popov, N.A. Smirnov, M.M. KhoKhlov, V.A. Savelichev, S.G. Dogaev, and J.M.Sokol. The computers were under construction using a modular principle based on the NC bus. The base block had a standard size of 5U Euromechanics (a power unit) and eighteen places for on-board modules. NC-03 software (the basis for all of them) included punched tape and disk OS, a library of standard programs, programming cross-systems on BESM-6 and ES EVM, an assembler, a system for debugging, a monitoring system, and a text editor. The designers of the base software included M.M. Khokhlov, V.S. Petrovsky, S.G. Dogaev, and N.S. Buslaeva.

3.3 Electronics NC-03T, NC-03D, and NC-03S

The NC-03T was designed during 1975-1976 in SVC. The MD was by D.I. Juditsky followed by J.E. Chicherin; the designers included V.E. Lukashov, V.S. Petrovsky, S.G. Dogaev, V.M. Yelagin, V.G. Sirenko, B.V. Shevkopljas, J.I. Borshchenko, V.V. Titov, JU.B. Terentjev, and L.M. Petrova. At the "Logica" factory at SVC, they began manufacturing an experimental batch from five computers. However, in connection with liquidation SVC and Logica, they completed the work in NII TT and Angstrem. It was a 16-bit computer that had one or two processors on the LSI K587; it supported up to 64K words of memory, four interrupt levels, and a system of commands. The NC-03 contained 190 commands. The computer was manufactured in Angstrem. Between 1976 and 1981, they released all 976 computers. The machine received the Gold medal at the Leipzig exhibition. The Electronics NC-03D computer was a more compact variant of the computer with the same basic characteristics, in the case 2U Euromechanics. Between 1978 and 1980, Angstrem manufactured and distributed 972 NC-03D computers. The Electronics NC-03S was a special NC-03D configuration for the "Electronics NC-32" system. Figure 11 illustrates these machines.

Fig. 11. Electronics NC-03T and NC-03D

3.4 Electronics NC-04T, -04U

In 1976, they designed the "Electronics NC-04T" computer based on the MPC K587. See Figure 12. The MD was N.M. Vorobjev and the designers included V.E. Lukashov, V.A. Savelichev, V.N. Shmigelsky, and V.A. Merkulov. The NC-03T had an expanded system of commands (up to 328), an arithmetic coprocessor, and developed systems of addressing and interruptions. Between 1980 and 1984, Angstrem had manufactured 1670 NC-04T machines.

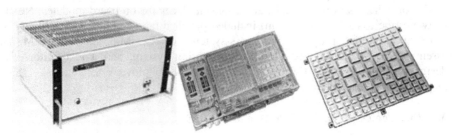

Fig. 12. Electronics NC-04T, NC-04U (I-04) and the ceramic printed-circuit-board

In 1977, they designed the NC-04U, which was a variant of the NC-04T for satellite onboard systems. The MD was V.A. Merkulov; the designers included A.M. Smagly, G.M. Alaev, A.E. Abramov, and E.V. Fedorova. The computer operated on a multilayered ceramic printed circuitboards with IC application in a micropackage (MPC N587). Angstrem manufactured 294 NC-04U machines between 1980 and1984.

3.5 Electronics NC-05T

The development of the "Electronics NC-05T" microcomputer began in 1979 as shown in Figure 13. The MD was M.D. Kornev; the designes included V.A. Savelichev, A.V. Bokarev, P.N. Kazantsev, J.M. Sokol, V.A. Khvorostov, V.I. Plotnikov, M.J. Gamorin, and ZH.A. Mamaev. They used the high-speed MPC series N1802, designed together with the NII ME, and built on multilayered ceramic boards, but in the typical case 5U for the NC computer. The NC-03 and the NC-04 differed in hardware realization of multiplication, division, a floating point in a 32-bit format of words, work in the mathematical space of addresses, and the protection of memory. Its speed was 1.0 million OPS. By the middle of 1981, it has had made and modified five samples of the NC-05T. However, at this time there were events that had fatally reflected its destiny.

Fig. 13. Electronics NC-05T

To 1981 in NII TT and in NPO NC change of generation of heads has come to the end. Send away veterans, founders NC possessed huge knowledge and experience of creation of complex radio-electronic systems. A change had come with the new generation, which has grown already in the Zelenograd center of microelectronics. They were experts in microelectronics, but not in computer facilities. The maximum authority on computer facilities for them was first deputy minister V.G. Kolesnikov, the active supporter of architecture PDP-11 of firm DEC. Between architecture of computer PDP-11 and NC they did not understand a difference, and to that architecture NC is younger than architecture PDP-11 for seven years (huge term in development of computer facilities), values have not given. Not having discussed with specialists and partners, management NPO NC has left with the offer to V.G. Kolesnikovu about cessation of work after architecture NC and transition to architecture PDP-11. It has agreed with readiness. Because of work on the architecture of NC, including the first version HЦ-05T, work stopped. Later other microcomputer with the same name, but already with architecture PDP--11/34 of firm DEC has been designed, then it has been renamed in NC-05D.

3.6 Electronics NC-31

In 1980, the NII TT had received the task to minister the reproduction of a system for programmed numerical control (PNC) for the firm Fanuc. The specialists of NII TT trained in SVC for independent design. They had suggested making a functional analogue based on the NC architecture, the MPC series K588, and the semi-customized LSI KR1801VP1-xxx. The minister has agreed, but had demanded full external conformity to analog machines. This resulted in the creation of the PNC "Electronics NC-31" computer as shown in Figure 14. The MD was J.E. Chicherin and the designers included V.N. Shmigelsky, V.N. Lukashov, J.B. Terentjev, J.I. Titov, V.S. Petrovsky, and I. Evdokimov. On set of parameters, the NC-31 did not concede to the best foreign models of that time.

Batch production of the NC-31 began in 1980 by Angstrem and then it transferred to the "Kvant" (Zelenograd) plant and to the «Diffuzion» (Smolensk) plant. The Angstrem and Kvant companies manufactured 3846 NC-31 computers.

Fig. 14. PNC "Electronics NC-31"

Machine tools for the NC-31 are still working, for more than twenty-five years.

3.7 Electronics NC-32

Since the DSC «Jurjuzsn» in the Pulkovo worked well, in 1978 MGA and the Ministry of Communications ordered a NII TT design for a multipurpose telegraph channel concentrator (TCC). By the end 1980, they designed the TCC "Electronics NC-32". The MD was N.A. Smirnov and the designers included M.D. Kornev, N.M.

Vorobjov, V.R. Gorovoj, P.P. Silantjev, V.A. Savelichev, A.I. Koekin, A.N.

Lavrenov, V.L. Glukhman, V.A. Merkulov, B.A. Mikhajlov, P.N. Kazantsev, I.P. Seleznev, V.I. Brikker, V.S. Petrovsky, and V.S. Travnitsky.

They designed the NC-32 was constructed based on the microcomputer NC-04T and the abonent station into its structure, based on the NC-03S machine. Its design supported special software. The NC-32 processed up to thirty-two telegraph channels with speeds of 50, 100, and 200 bps.

The first complete NC-32 system was installed at the Central telegraph in Moscow,

Fig. 15. Abonent station TCC "Electronics NC-32"

where it replaced three-hundred operators and had payback period of nine months. The first 730 NC-32 machines of various configurations have been led by Angstrem that equipped all (nearby 200) republic and regional telegraphs of the USSR including many airports. Further batch production the NC-32 transferred to the Cherkassk factory for the telegraph equipment.

3.8   Electronics «Tonus NC-01»

In 1980, the NII TT was based on the MPC K587 and it minimized the NC architecture. They designed a portable medical complex «Electronics Tonus NC-01», with N.N. Zubov as its MD. See Figure 16. Its purpose was an automatic estimation of working capacity, psychological activity, and the forecast of efficiency of professional work of an operator (e.g. pilot, driver, cosmonaut, sportsman, dispatcher). The experimental batch for the Tonus NC-01 made fifteen complete sets had passed pre-production operation in the different medical research centers. In 1982, however, without warning they stopped work on medical subjects in NII TT.

Fig. 16. Electronics Tonus NC-01

3.9   Electronics NC-80T

In 1980, the NII TT designed the n-MOΠ 16-bit single-chip computer with the NC architecture; the chip was the K1801VE1 as shown in Figure 17. The MD was V.L. Dshkhunjan and the designers included P.R. Mashevich, P.M. Gafarov, S.S. Kovalenko, A.A. Ryzhov, V.P. Gorsky, and A.N. Surkov.

The K1801BE1 was a 16-bit computer with a possibility of processing 1, 8, 16, and 32-bit data. It had an addressable space of 64K words (128K byte), a resident (on a chip) RAM of 128x16 bits, a ROM of 1024Kx16 bits, and system of commands of the NC-03.

Because of the limitation of the number of pins in the LSI, they applied a variant of the NC bus with a combined line of the address and data. For peripheral devices, it completely corresponded to the Q-BUS bus of LSI-11 microcomputer of the DEC firm, but had different (up to four) microprocessors. The bus had received the name "The Main parallel interface (MPI)" and it is legalized by the OST 11.305.903-80 and GOST 26765.51-86 standards. The K1801VE1 contained the microprocessor, RAM, ROM, timers, ports for input-output, and an on-bus MPI output.

Fig. 17. K1801VE1 in the package (full-scale) and it topology

Fig. 18. Electronics NC-8001

3.10 Electronics NC-8001

In the beginning of 1981, they designed the "Electronics NC-8001" onboard computer, based on the K1801BE1. See Figure 18. The MD was V.L. Dshhunjan and the designers included N.G. Karpinsky, A.I. Polovenjuk, N.I. Trofimova, and I.O. Lozovoj. It could process 1, 8, 16, and 32-bit data with speeds up to 500,000 OPS. Its structure included RAM and ROM on 32K byte, a 16-digit timer, 32 programmed lines of input/output, and ports for the displays and printers. The computer is mounted on a printed circuit-board in the size 180x300 mm with sockets from two connectors: one on the MPI bus and another for external ports.

3.11 Electronics NC-8020

In 1981, they designed a multi-board small-sized (less the block 2U or 5U) the computer based on the NC-8001. Two-board and eight-board blocks are designed for installation on NC8001 and peripheral modules. In the first model of computer, NC-8020 (see Figure 19) was two modules: the NC-8001 and the KSPK for connection of peripheral devices.

3.12 Electronics NC-8010

In May of 1981, the designed via the NII TT and based on the K1801BE1 by the computer of individual using "Electronics NC-8010", that was program-compatible with the NC-03T. See Figure 20. The MD was V.L. Dshhunjan and the designers included A.N. Polosin, N.G. Karpinsky, A.I. Polovenjuk, O.L. Semichasnov, B.G. Beketov, A.R. Razvjaznev, and I.O. Lozovoj. It was the first personal computer in the USSR and it was constructed completely on homemade microcircuits. It used homemade architecture that was program-compatible to a homemade family of "Electronics NC" microcomputer. The NC-8010 was a dual-processor (two K1801VE1 for the central processor and the processor of input-output) system with two programmed ports (64 communication lines). As a video monitor and the external storage, they used a household TV (512x256 pixels) and a compact-cassette tape recorder. Structurally, the NC-8010 was built in the casing of a keyboard and it was intended for the decision in a dialogue mode of scientific, engineering, educational, and problems.

Fig. 19. Electronics NC-8020   **Fig. 20.** Electronics NC-8010

The NC-8001, NC-8010, and NC-8020 operated normally. Nevertheless, at that time there was a DEC-revolution as described above. The NC architecture appeared under an interdiction. The works above (NC-05T, K1801VE1, NC-8001, NC-8010 and NC-8020) stopped. Certainly, it was the best microcomputer in the country at that time, comparable to the best foreign samples.

Fund Collection: "Electronic Digital Computing Machines" at the Polytechnic Museum

Marina Smolevitskaya

Scientific Researcher, Computer Collection Curator
Polytechnic Museum, Moscow, Russia
smol@polymus.ru

Abstract. The Polytechnic Museum began to collect calculating devices and computing machines in the first years of its formation in the 1860s.. Today, the Museum has the Fund Collection "Electronic Digital Computing Machines", which consists from seven systematic collections and eleven personal funds of Russian scientists. There are about three hundred objects and over sixteen hundred documentary, printed, and graphical items today. All four generations of electronic digital computing machines are presented in the museum. In addition, the museum created eleven personal funds of Russian scientists who devoted their activity to computer science. The museum has opened these funds, which include collections from S. Lebedev, I. Bruk, B. Rameev, V. Glushkov and others. It is very important to point out that this fund collection is the only one of such variety and size in Russia.

Keywords: Polytechnic Museum, collection, personal funds, electronic digital computing machines.

1 Introduction

The idea of establishing a Polytechnic Museum first took place in the 1860s. The country had been going through the reforming epoch of Alexander II. The foundation of a national capitalism had also been forming, thus, new knowledge and technological ideas were greatly required.

In 1864, the "Emperor society of amateurs for natural sciences anthropology and ethnography" (ESANSAE) had emerged. The main task undertaken by the scientists who had joined the society was their assistance to scientific progress and the dissemination of natural and scientific knowledge. As an educational complex, the members of the emperor society founded the first library, known as the Central Polytechnic Library. Later in 1872, they established the General Educational Museum of Applied Sciences, known to us today as the Polytechnic Museum.

Some of the exhibits of All-Russian Polytechnic Exhibition of 1872 were devoted to the bicentennial of Peter the Great; these exhibits became the foundation for the future museum. The computer engineering department was one of those exhibits and had certainly attracted everyone's attention. The simplest devices for counting and the up-to-date computers represented the history of their progress. The Polytechnic

J. Impagliazzo and E. Proydakov (Eds.): SoRuCom 2006, IFIP AICT 357, pp. 187–193, 2011.

Museum began to collect calculating devices and computing machines in the first years of its formation. Today, the Museum has the Fund Collection "Electronic Digital Computing Machines", which consists from seven systematic collections and eleven personal funds of Russian scientists.

2 The Fund Collection

The fund collection of "Electronic Digital Computing Machines" (EDCM) was formed in the 1960s. There are about three hundred objects and over sixteen hundred documentary, printed, and graphical items today. In addition, the museum created eleven personal funds of Russian scientists who devoted their activity to computer science. The museum has opened these funds, which include collections from S. Lebedev, I. Bruk, B. Rameev, V. Glushkov, U. Bazilevskiy, N. Matjukhin, M. Kartsev, A. Kitov, N. Brousentsov, V. Petrov, and V. Burtsev. It is very important to point out that this fund collection is the only one of such variety and size in Russia.

The first electronic digital computing machines appeared in our country in 1951 and allowed scientists to solve difficult scientific and technical tasks. They were the Small Electronic Calculating Machine (MESM), developed under leadership of academician Sergei Lebedev, and the Automatic Digital Computer (M-1), developed under leadership of Isaak Bruk. N. Matjukhin and M. Kartsev were among the developers of the M-1 computer and later, they created computer engineering schools of their own. The documentary materials about these machines and its developers appear in the halls of computer engineering. The original report of the M-1 Automatic Digital Computer, developed in the Laboratory of Electro-systems at the Institute of Energy of the USSR Academy of Sciences, is one of the most interesting documents in our department.

In 1948, Isaak Bruk together with Bashir Rameev received the first author's certificate of the Automatic Digital Computing Machine in Russia. The museum is the custodian of this certificate. Later, B. Rameev created the "Ural" family of computers. One can see the Small Automatic Electronic Digital Computing Machine, the "Ural-1", in the Polytechnic Museum exhibition. Some museum objects of computer science have obtained the status of "Relic of science and technology"; as such, they are under protection of the museum and the state. The Small Automatic Electronic Digital Computing Machine "Ural-1", some units of the first Soviet serial computer "Strela" and other museum objects have designated such status. There are several units of the first Soviet serial computer "Strela", developed in 1952 by the Special Engineering Bureau SKB-245 in the EDCM Fund collection of Polytechnic Museum. These are the fragments of the Control Pulte, several processor blocks realized on the vacuum-valves, the six cathode-ray tubes (elements of quick-access storage), and wide ferromagnetic tape used as an external information carrier.

3 Generation of Machines

Usually all electronic digital computing machines are divided into four generations. The Small Automatic Electronic Digital Computing Machine "Ural-1" presents the first generation of machines in the museum. The processor of these machines was realized on electronic tubes, and operative memory was realized on magnetic drum or cathode-ray tubes.

Then the second generation of machines is represented by electronic digital computing machine "Razdan-3". They built the processor using semiconductors and operative memory on ferrite cores. There were several ferrite cubes in one machine. There are many matrixes of ferrite cores inside such cube. We can see how these devices worked on the demonstration model.

The museum collects and keeps the various types of memory on ferrite cores. For example, it contains the Ferrite Cube of the Operative Memory of the Electronic Computing Machine (ECM) M-4, developed under leadership of M. Kartsev. The capacitor-type ROM block of the ECM M10 is a very interesting object, which was designed in the Scientific Research Institute of Computing Complexes also under the leadership of M. Kartsev.

The unified system of electronic computing machines on integrated circuits represents the third generation, developed in the USSR at the beginning of the 1970s in cooperation with the socialist countries. It represents a family of software compatible machines with different productivities that build on the unified elemental and constructive base with a unified structure and a unified set of peripheral units. The processor and the operative memory were mounted on integrated circuits.

We can see the Computing Center with the third generation machines through the model of the Electronic Computing Machine ES-1050 and some original units of this machine. We can see integrated circuits, which made the peculiar revolution in computing science, on the boards of the operative memory of the ES-3222. Plotters were used widely with these machines for the first time.

The processor and operative memory of the fourth generation of electronic computing machines appeared on very large-scale integrated circuits. We demonstrate the functioning of one of the first such Soviet computers – the Microprocessor Laboratory "MikroLab КР580ИК80" in the museum.

A remarkable exclusion is the experience of creating the ternary computers "Setun" and "Setun-70"at Moscow State University. The experience convincingly confirms practical preferences of ternary digital technique. N. Brousentsov initiated the design of small digital computing machine "Setun" in 1956. (Note that Setun is the little river that flows into the river "Moscow" near the University.) The Setun was a small, inexpensive computer that was simple to use and to service for schools, research laboratories, offices, and manufacture control. Fast miniature ferrite cores and semiconductor diodes were used as the element base for this machine. Simplicity, economy, and elegance of computer architecture are the direct and practically very important consequences of ternary machines. The computer "Setun" has the status of "Relic of science and technology".

4 Special Machines and Recognition

There is a unique computer for spaceship use called the "Argon-16" that can be seen only in this museum. It contained a synchronous computer system with triple redundancy and majorization carried out on per unit base with eight levels. It consists of three computers with data channels and a set of interfaces to the control system. The instruction set is specially designed for control tasks. I/O operations combine with the calculation process.

Since 1975, the "Argon-16" computer became the basic component of control systems of "Soyuz" spaceship, the "Progress" transport ships, and orbital space stations "Salute", "Almaz", and "Mir". Exclusive reliability had provided long usage for it. The total output for these machines is 380. No failure of the system was noted during its twenty-five years of operation when working in control systems. It is unrivalled among space computers by production volume. The specialized computers for aviation are presented in the Museum exposition.

In 2005, the museum received the "El'brus 3.1" computer system, developed at the Institute of Precision Mechanics and Computer Technology. It also received one processor block and one operative memory block of the "Electronica SS BIS-1" super computer, created under the leadership of academician V. Mel'nikov at the Cybernetics Problems Institute and the "Del'ta" Scientific Research Institute. In addition, the museum actively collects Russian and foreign personal computers.

We have on exhibition in the museum halls devoted to Soviet scientists and engineers who worked in computing science. We can see documentary, printings, and graphic materials of scientific, official, and biographic activity of these people in this exhibition.

In addition, there are materials about international recognition of Russian scientists in computer science. The International Computer Society of the Institute of Electrical and Electronic Engineers awards scientists of different countries the title "Computer Pioneer". Russian scientists S. Lebedev, A. Lyapunov, and V.Glushkov received this title in 1996. These diplomas and large bronze medals were given to the children of these scientists.

Since 1994, the Department of Computer Engineering and the Automata Department of the Russian Academy of Science awards a premium, named after S. Lebedev, for successes in area of computing systems development. This museum keeps copies of the diplomas awarded by Russian Academy of Sciences to Russian scientists.

5   Curator Activity and Scientific Research

It is well known that every Museum begins from an object that was donated or purchased. After that, this object is accepted for temporary registration. Therefore, a scientific researcher composes a detailed report for the special museum commission that decides whether to keep this object.

The collection curators write scientific conceptions of collections and they compile conception programs of them. The museum has the following scientific conceptions among others:

- o Electronic Digital Computing Machines,
- o Specialized Electronic Digital Computing Machines for military applications,
- o External Data Medium,
- o Electromechanical Calculating Machines,
- o Punched-Card Machines, and
- o Simple Calculating Analogue Instruments and Mechanisms.

The collection scientific conception determines the collection object and the collection function. In this document, curators compile the historic information and the objects completing and selection principles. Then they write the collection structure and composition. This document determines the curator activity in collection forming over many years.

One of the important types of the curator scientific research is the scientific description of the most important museum objects: writing the "Scientific Passport of the Museum Object". In this document, we try to gather the maximum information about the object and to interpret this information for the purpose of historical, scientific and museum significance.

The "Passport of the Museum Object" document is compiled for the most important and valuable objects of the collection. The document has thirty-seven information fields. It contains the following main parts: the "object description", the "operation principles", the "technique parameters", the "museum significance", "literature", and the "appendix". In the "Appendix" we can place information such as the designer biography of the object and the patents obtained on this object. All information in the "Scientific Passport of the Museum Object" must exist to confirm references to the information sources. Such sources are the self-object, literature, states in periodical publications, technical documentation, archives, and information from specialists.

One of the most important purposes of the Polytechnic Museum is to discover and to select the relics of science and engineering, to describe research, classify and systematize them, to take care of them (restoration, conservation, and protection), to introduce them into scientific use, and to popularize them.

A "relic of science and technology" is the material object, directly or indirectly connected with main stages of science and engineering development. This requires to keep the relic in conformity with its social and scientific significance and to use it in the general cultural system. Special information cards are compiled for such objects. Then an expert committee of the Polytechnic Museum, appointed from the Association of Scientific and Technology Museums of Russian Nationality Committee of ICOM, confers the status "relic of science and technology" if warranted.

Eight objects have achieved this status in the Fund Collection "Electronic Digital Computing Machines". These are

- The Small Automatic Electronic Digital Computing Machine "Ural-1",
- The Electronic- rays Tube and The Ferromagnetic Tape of the first Soviet serial computer "Strela",
- The Control Pulte of The Small EDCM "Setun",
- The ECM "MIR-2",
- The Ferrite Cube of The Magnetic Operative Memory,
- The Magnetic Drum of ECM "Minsk-32",
- The Unit for abonent's linking of the ECM M13, and
- The Capacitor Type ROM Blok of ECM M10.

Work on discovering "Relics of science and technology" will continue indefinitely.

The final stage of the curator scientific research is compiling the scientific catalogue on the fund collection. The catalogues "Mechanical Musical and Curious Automata", "Telegraph Sets", and other documents are published under the "Cultural Russia Heritage" program.

Today, the Fund Collection "Electronic Digital Computing Machines" is actively being enriched and explored, so it is too early to compile a catalogue on this theme. Together with the academic institutes, trade research-production and exploratory centers, universities, various departments, and public scientific and technological organizations, the museum conducts scientific conferences, readings, round table discussions, and meetings devoted to distinguished Russian scientists and engineers, and important dates in history of science and technology.

The Polytechnic Museum has printed the albums "Relics of Science and Technology", books with speeches from the participants of polytechnic readingsthat are part of the museum. For example, it hosted "Cybernetics: Expectations and Results" and "Specialized Electronic Digital Computing Machines for the Army ".

Within its program called "Remarkable Engineering Projects of Russia", the museum printed books in which showed the results of scientific research of creative heritage of outstanding Russian scientists in area of computer engineering. For example, it published the article "Pioneer of Soviet Computing Industry: Sergey Lebedev" and "Creative Heritage of B. Rameeva: One of the Founders of Domestic Computing Technology".

6 Some Reflections

Today, the Polytechnic Museum is rightfully considered the main museum in Russia, showing the country's history of science and technology. As a scientific and methodic center of the museum management studies, it fulfills an entire set of important tasks.

The museum brings to light and shows the collections of other scientific and technical museums to further a fair preservation of national heritage. It is guiding a professional skill mastering in the work of the museums of technical profile. The museum coordinates the activities on finding, custodianship, and bringing to cultural circulation the most valuable, rare objects of science and technology from other museum collections of the country. It also assists in sharing the experience of foreign and home specialists with the scientific and technical museums.

As the leading museum of science and technological history, the Polytechnic Museum renders the methodic and practical help in establishing museums. In 1987, the Polytechnic Museum took part in the foundation of the Association of Scientific and Technical Museums under Russian Committee of ICOM (International Council of Museums). Today, the museum is a scientific organization and methodic center for the Association. Since the 1970s, the Polytechnic Museum is a Member of the International Council of Museums.

Students from several Moscow universities have internships in the museum. The museum suggests some themes and works that they can do. The main themes are:

1. Imitation of functioning different calculators and computing machines with help of new information technologies;
2. Creating slide shows on computing history through PowerPoint presentations;
3. Repairing and restoration the objects of the EDCM collections.

Students of Moscow State Academy of Instrument-making and Informatics created some animated models of objects from the calculating instruments collection using a modern professional program called Macromedia Studio MX. Chebushev's arithmometer and S. Dgewons' logical machine are among them.

An example of student work on the second theme is the slide show titled "The First Personal Computer: The Altair". A student of the Moscow College of Information Technologies created it (№1533). Now for any categories of visitors it is possible to demonstrate with the PC the following slide shows developed by the author of report with the aid of the program PowerPoint:

o "Charles Babbage and his Computers";
o "Ada Lovelace - the First Programmer in the World";
o "Search of Versions in the Composition of Fairy Tales";
o "Anthem to Artificial Intellect".

The third topic is the most difficult. Unfortunately, it is practically impossible to restore the first electronic digital machines for demonstration of their work. However, with the aid of modern computers it is possible to show the operation of the separate devices. One of such complex demonstration was created on the base of plotter "US-7051M" and the personal computer, which works in the DOS medium. This plotter worked under the control of a special block in composition of United System computers. We do not have such a control block, but the student of Moscow Institute of Electronic and Mathematics developed a new interface between this plotter and personal computer on modern microchips (integrated circuits). The control programs, written in the algorithmic language C++, and the demonstration programs allow drawing of images chosen by the user from the computer library, for example, the logo of Polytechnic Museum. It is important to note that in the created demonstration complex in the base of plotter US-7051 and personal computer partially the history of appearance and development of the algorithmic languages is reflected and remains the same.

With the aid of modern multimedia computer, it is also possible to listen to computer music: from the first solo-voice melodies "Uralskie napevy" of R. Zaripov, to the polyphonic compositions of A. Stepanov (played on the first computing machines in the 1960s). They were rewritten from old recording tapes and enumerated before modern plays. They are written for the computer and the usual musical tools such as "Dialog of computer and violoncello".

In 2005, the museum began to carry out the work according to digitization of video films from the scientific-auxiliary fund of the museum. Five films are already in digital format. For example, the museum digitized "Academician Lebedev", "Machine Geometry and Graphics" and "Curved Surfaces in Automatization System". The museum plans to continue this work.

All scientific research and knowledge about the museum objects are used for enlightening, educating, and developing social-cultural activities in the museum. During the excursions "How People Learn to Calculate" and "From the Plum Stones to the Clockwork Calculators", the youngest schoolchildren can calculate on devices such as the Asian abacus, Russian frame-wooden counters, and arithmometers.

For the older school children, students from the professional schools and high schools, the museum holds overview excursions. These include "From the Abacus to the Modern Computer", "The world of EDCM", and some excursions that reveal more completely themes such as "Artificial Intelligence of Computer" and the "Time Overtake Project" (devoted to the Charles Babbage Analytic Engine and Ada Lovelace), and the "Information Revolution in the Civilization History". During the autumn, spring, and winter holidays the museum holds short excursions for random visitors, usually parents with their children.

An Open Adaptive Virtual Museum of Informatics History in Siberia

Victor N. Kasyanov

Head of Laboratory of Program Construction and Optimization
A.P. Ershov Institute of Informatics Systems
Lavrentiev pr. 6, Novosibirsk, 630090, Russia
kvn@iis.nsk.su

Abstract. As with the history of any other science, the history of informatics (computer science), is an important and inseparable part of this science. In the paper, the Siberian Virtual Museum (SVM) project that is under development at the A.P. Ershov Institute of Informatics Systems and aimed at development of an open adaptive virtual museum of informatics history in Siberia is described. The conception of an open adaptive virtual museum and the objectives of the SVM project are discussed. The architecture and users of the SVM museum are considered.. The adaptive virtual museum SVM is intended to be accessible annals of Siberian informatics history, which can be written by active users.

Keywords: Digital museum, virtual museum, open virtual museum, open adaptive virtual museum, informatics history in Siberia.

1 Introduction

With the advent of the digital age and the web, museums and cultural heritage institutions began rethinking their roles. An increasing number of museums make the decision to maintain a website (a *digital museum*) to provide useful information and attract new visitors. The advantage of digital museums is clear. The visitors of a digital museum can enjoy cultural relicts without a restriction of time and place, and complete safety of cultural relicts is guaranteed. The visitors have the opportunity to see precious cultural relicts that cannot be exhibited in a conventional way for reasons of safety or security. Furthermore, with the help of multimedia interaction, the visitors can even "touch" or "manipulate" the objects, which would be important for professionals. To extend access to networked browsing and querying, issues arise of collaboration between institutions and of standards for access [16].

Along with the "classic" digital museums which are websites of real museums, there are so-called virtual museums [7, 15, 17 - 21]. A *virtual museum* in this context refers to a repository of digital cultural and scientific resources that can be accessed and used anytime, anywhere via the internet. This means it is a website (a digital museum) that can but does not have to have any corresponding real museum and contains virtual exhibits being multimedia digital representations of any artifacts without a restriction of their nature or current state. For example, a virtual exhibit can present a painting being lost, a painter, a school of painting, or an art event.

J. Impagliazzo and E. Proydakov (Eds.): SoRuCom 2006, IFIP AICT 357, pp. 194–200, 2011.

From the viewpoint of museum visitors, a real museum is an environment for excursions and expositions. On the other hand, museums are cultural heritage institutions intended to support collecting, research, making catalogues and exhibiting artifacts; however, museum visitors cannot take part in this important museum work. We believe that virtual digital museums can be "open museums" that allow extending this museum work to wide range of virtual museum users. We assume that it is useful that a museum user can propose a presentation of some real artifact as a virtual exhibit to an open virtual museum. In addition, an open virtual museum may also have facility to supply exhibits with author descriptions, to offer guided tours around the museum, and to make a curatorial exposition. These possibilities are very important for modern history museums. An *open virtual museum* [10] is a hypermedia system intended to be both an accessible repository for artifact collections and a cultural heritage institution supporting the collective work of many people, which are interested in collecting, annotating, organizing, research, making catalogues, and exhibiting these artifacts.

Adaptive hypermedia is an alternative to the traditional "one-size-fits-all" approach in the development of hypermedia systems [2]. Adaptive hypermedia systems build a model of the goals, preferences and knowledge of each individual user, and use this model throughout the interaction with the user to adapt to the needs of that user.

Open adaptive virtual museums can support the accessibility and active use of digital cultural and scientific resources for everyone without a restriction of time and place. They can bring several benefits described as follows.

o   Collective work of many people who are interested in collecting, annotating, organizing, research, making catalogues and exhibiting any artifacts;
o   Virtual exhibitions that cannot be organized otherwise, e.g. a comprehensive exhibition of an artist whose works are distributed all over the world in public and private collections;
o   Private collections and artifacts can be made available for public, taking into account various levels of anonymity for the owner – anonymous, semi-anonymous (i.e. available for discussions under a nickname), non-anonymous, available for a visit, etc;
o   Exhibitions on demand can be organized for visitors;
o   Adaptive guided tours can take place for each individual visitor taking into account her interests, preferences and constraints (like time).

In the paper, the Siberian Virtual Museum (SVM) project of the open adaptive virtual museum of informatics history in Siberia is described [10 –12, 19]. The paper's structure is as follows. Section 2 presents the main objectives of the SVM project. The architecture of the SVM museum is briefly considered in Section 3. Section 4 describes the users of the SVM system. Section 5 provides a conclusion.

2   Objectives of the SVM Project

The history of informatics (or computer science), as with the history of any other science, is an important and inseparable part of computing. Teaching the history of computing has become a part of the computing curriculum of many Western universities. A special IFIP Joint Task Group has published a comprehensive report containing a number of valuable methodological instructions [8].

At the same time, informatics history of Eastern Europe and the USSR was practically unknown in Western Europe, although some works on this problem have been published [5, 6]. In 1996, the IEEE Computer Society, in connection with the 50th anniversary of its foundation, presented the Computer Pioneer Award to sixteen scientists from Central and Eastern Europe countries, including an outstanding Russian scientist and academician Alexej A. Lyapunov, who "developed the first theory of operator methods for abstract programming and founded Soviet cybernetics and programming" [4].

In Siberia, research in programming started after Alexej A. Lyapunov and his disciple Andrei P. Ershov had arrived to the Novosibirsk Academgorodok (in the early 1960s). Academician A.P. Ershov and his followers have founded the Siberian school of programming and informatics; this was the third one in the USSR, after Moscow and Kiev. Now, many years after its founder A.P. Ershov died [1], it continues to pay an important role in spite of all the difficulties endured by the Russian science and education. This gives us an opportunity to investigate independently the formation and development of informatics in Siberia, namely, in the Novosibirsk Scientific Centre, against the Russian and world scenes.

For fifty years, informatics had developed very intensively in Siberia, but by now some active participators and eyewitnesses of its development have died, many facts have been lost, and some are still unknown. So, it is very important to have an open virtual museum of informatics history in Siberia. It is believed that this museum can take the form of accessible annals of the Siberian computer science history, which can be written by active users.

Most of the museums presented on the www now are traditional hypermedia systems and provide the same information and navigation methods to all users. At the same time, a few papers discuss how the design of websites containing museum information could be improved to take into account various needs of different visitors. Paper [13] describes the use of natural language generation technology in the construction of personalized virtual electronic catalogues for a variety of domains such as digital museums, encyclopedia, and tourist guides. Paper [3] describes an evolution of the intelligent labeling explorer (ILEX), a system that dynamically generates personalized text labels for exhibits in a museum's jewelry gallery. Paper [14] describes a portable system (an adaptive museum guide implemented on a hand-held computer), which provides a visitor with personalized navigation help and information about visited objects.

The SVM is intended for use by different categories of users, and it is very important that museum users with different preferences, goals, knowledge, and interests may obtain different information and may use different ways of navigation. Therefore, we give a particular attention to adaptation problems in our project [10 - 12].

3 Architecture of SVM

Currently, databases (DBs) of the web-based SVM museum provide storage and processing of the information about the following objects: publications, archive documents, projects, data about scientists in informatics, scientific teams, various events concerning informatics history, conferences, and computers. All the above objects are the *exhibits* of the virtual museum. Every exhibit has the following main attributes: a Unique Universal Identifier (UUID) of an object, a name, sometimes a date, a brief description (or an annotation), a full description (or a file), a name of a person who presented this exhibit, the date of its addition, the possibility and permission of its modification and participation in exhibitions.

3.1 Tours and Exhibits

A set of exhibits united according to the thematic, chronological, or typological criteria can be represented as an *exhibition* or a *tour* (or an *excursion*). Both an exhibition and a tour have the following attributes: UUID, a name, a name of a person who created it, brief description, and reference(s) to the file(s) representing it contents. Main differences between an exhibition and a tour are the following.

o   A tour is composed of one section (a file), while an exhibition can consist of several sections (exhibitions or sub-exhibitions);
o   A tour is a story about the museum (elapsing in the time) followed by demonstration of its exhibits in a definite order. A tour, for example, may be a clip or a presentation for MS PowerPoint and may not only be in an online mode but sometimes even exist offline. In contrast to a tour, an exhibition consists of exhibits that a visitor is looking at by himself and only online. Usually, several ways of navigation, including a free movement among exhibits, are available.

All exhibitions (and tours) are divided into *permanent* and *temporary* ones. A hall of *exhibitions* and a hall of *tours* are designed as accessible to all users of the museum.

3.2 Halls and Some Nomenclature

There are also restricted halls in the museum: the library, the archive, the chronicle, the halls devoted to scientists in informatics, scientific teams, projects, computers, conferences, the hall of new exhibits and the hall of preparation of exhibitions and excursions. These halls are accessible only to registered users of the museum (see below).

The *library* consists of books, articles and so on. In addition to the general exhibit attributes, each library exhibit has a list of authors and other attributes.

The *archive* consists of text, graphic, audio, and video materials.

The *chronicle* of events contains a description of the most remarkable events of informatics history in Siberia.

The *hall of scientists in informatics* presents information about the prominent scientists in informatics. In addition to the general information, it provides the following data about scientists: their education, scientific degrees, titles and posts, scientific interests, the text of the biography, photos, the main publications and projects.

The *hall of scientific teams* presents information about groups, laboratories and institutes. Along with the general attributes, each team has its address, etc.

The *hall of projects* provides information about projects in informatics including the dates of its beginning and finishing.

The *hall of computers* shows computers, which were used and created in the Siberian Division of the Russian Academy of Sciences. In addition, each exhibit has the name of the designer, and the photo.

The *hall of conferences* contains the following information about each scientific event: where and when it was held, its status, and the general exhibit information.

New entries to the museum (adding by users) are placed in the *hall of new exhibits*.

Exhibitions and excursions created by users of the museum are being composed in the *hall of preparation of exhibitions and excursions*.

4   Users of SVM

All users of the SVM web-based museum are divided in two main categories: unregistered users (*visitors*) and registered ones (*specialists*) with different access level to information resources.

Visitors have access only to the part of museum information that is opened for public access (for example, in the form of excursions and exhibitions). In this case, all resources are accessible only for review and search. Visitors are divided in two subcategories depending on their knowledge level of subject domain: *beginners* and *experts*. Beginners have an opportunity to look only at tours, and experts can also look at exhibitions and electronic conferences of users.

Specialists have access to reviewing all information resources of our museum, including restricted halls closed for public access; they can also take part in electronic conferences and write in a visitors' book.

All specialists are divided in two main groups depending on their level of access to resources: a group of *simple specialists* working only in the hall of new exhibits and a group of *museum employees*.

Volunteers, tour guides, and *exhibition curator/designers* are selected from a group of simple specialists. Volunteers have permissions to add new exhibits of any type. Tour guides may create their own tours, and curators/designers the exhibitions. Objects made up or added by them are initially placed in the hall of new exhibits, and then managers of the corresponding resources (for example, chief tour guide or head of exhibitions) decide whether to include them in the museum's resource. Volunteers, tour guides, and exhibition designers have no permissions to modify the museum databases.

A group of museum employees can be presented as a hierarchical structure, with a *director* (or the senior manager) at its very top. He has full authority to administrate the museum DBs, including DB of museum users.

The second level of the hierarchy consists of the *managers* (or *administrators*) of the corresponding museum resources. They are appointed by a director such as the head of exhibitions, the chief tour guide, the chief librarian, the chief archivist, the chief chronologist, the chief biographer, the chief expert on scientific teams, the chief planner, the chief engineer, or the chief secretary. Resource managers (administrators) have full authority to administrate DBs of the corresponding resource types. They also control specialists working with DBs of corresponding types of resources.

The third level of the hierarchical structure includes museum employees appointed by the managers of the corresponding types of resources such as librarians, archivists, historians, biographers, experts on scientific teams, planners, engineers, and secretaries. They have limited rights to change DBs of the corresponding resource types.

5 Conclusion

We considered the conception of an open adaptive virtual museum that supports the accessibility and active use of digital cultural and scientific resources for everyone without a restriction of time and place. We also presented the SVM project of an open adaptive virtual museum of informatics history in Siberia.

The main purpose of creating SVM is to save historical and cultural heritage, the history of creation and development of computer science in Siberia. The SVM is also intended to provide a free common access to pages of the true history of computer science in Siberia, and therefore to increase cultural and educational level of people. It can be used as accessible annals of Siberian computer science history, which can be written by active users.

The solutions here considered can be used in the development of other virtual museums related to modern history or needed in collective work of people from different places. They can also be useful in the development of digital websites of real museums to support integration of knowledge and skills of museum workers from different museums. In particular, they were used in the DAVON project, proposal of which was under development to submit to the Sixth Framework Program, Call 5. The DAVON project aims to develop methods and tools that support open adaptive virtual museums of art and science history in Europe.

Acknowledgment. The author is thankful to all his colleagues, taking part in research connected with elaboration of the SVM. The SVM project is based on informatics history pages of the web-system SIMICS [10] and is supported in part by the Russian Foundation for the Humanities (Grant No. 02-05-12010).

References

[1] Bjorner, D., Kotov, V.: Images of Programming. Dedicated to the Memory of A.P. Ershov. North-Holland, Amsterdam (1991)
[2] Brusilovsky, P.: Adaptive hypermedia. User Modelling and User-Adapted Interaction 11(1), 87–110 (2001)
[3] Cox, R., O'Donnell, M., Oberlander, J.: Dynamic versus static hypermedia in museum education: an evaluation of ILEX, the intelligent labelling explorer. In: Proceedings of the 9th International Conference on Artificial Intelligence and Education, Le Mans, pp. 181–188 (1999)
[4] CS Recognizes Pioneers in Central and Eastern Europe. IEEE Computer (6), 79–84 (1998)

[5] Ershov, A.P.: A history of computing in the USSR. Datamation 21(9), 80–88 (1975)

[6] Ershov, A.P., Shura-Bura, M.R.: The early development of programming in the USSR. In: A History of Computing in the Twentieth Century, pp. 137–196. Acad. Press, New York (1980)

[7] European Virtual Computer Museum. Development of Computer Science and Technologies in Ukraine, http://www.icfcst.kiev.ua/museum/

[8] Impagliazzo, J., Campbell-Kelly, M., Davies, G., Lee, J.A.N., Williams, M.R.: History in the Computing Curriculum. IEEE Annals of the History of Computing 21(1), 4–16 (1999)

[9] Kasyanov, V.N.: SIMICS: information system on informatics history. In: Proceedings of International Conference on Educational Uses of Information and Communication Technologies.16th IFIP World Computer Congress, PHEI, Beijing (2000)

[10] Kasyanov, V.: SVM - Siberian virtual museum of informatics history. In: Innovation and the Knowledge Economy: Issues, Applications, Case Studies, Part 2, pp. 1014–1021. IOS Press, Amsterdam (2005)

[11] Kasyanov, V.N., Nesgovorova, G.P., Volyanskaya, T.A.: Adaptive hypermedia and its application to development of virtual museum of Siberian informatics history. In: Proceedings of International Conference PSI 2003, Novosibirsk, pp. 10–12 (2003) (In Russian)

[12] Kasyanov, V.N., Nesgovorova, G.P., Volyanskaya, T.A.: Virtual museum of informatics history in Siberia. Problems of Programming (4), 82–91 (2003) (In Russian)

[13] Milosavljevic, M.: Electronic Commerce via Personalised Virtual Electronic Catalogues. In: Proc. of 2nd Annual CollECTeR Workshop on Electronic Commerce (CollECTeR 1998), Sydney (1998), http://www.dynamicmultimedia.com.au/papers/collecter98/

[14] Not, E., Petrelli, D., Sarini, M., Stock, O., Strapparava, C., Zancanaro, M.: Hypernavigation in the physical space: adapting presentation to the user and to the situational context. New Review of Multimedia and Hypermedia 4, 33–45 (1998)

[15] On-line Museum of Computer History. Project of MGTU, http://museum.iu4.bmstu.ru/project.shtml

[16] The CIMI Profile Release 1.0H A Z39.50 Profile for Cultural Heritage Information, http://www.cimi.org/old_site/documents/HarmonizedProfile/HarmonProfile1.htm

[17] The Russian Virtual Computer Museum. Project of Eduard Projdakov, http://www.computer-museum.ru/

[18] The Virtual Museum of Manchester Computing, http://www.computer50.org/kgill/

[19] Virtual Museum of Informatics History in Siberia, http://pco.iis.nsk.su/svm/

[20] Virtual School Museum of Computer Science, http://schools.keldysh.ru/sch444/MUSEUM/

[21] Virtuelles Museums der Informatik, http://www.fbi.fh-darmstadt.de/~vmi/

The History of Computers and Computing in Virtual Museums

Yuri Polak

Chair for Social Informatics, Moscow State University
and Russian Academy of Science, Moscow, Russia
polak@cemi.rssi.ru

Abstract. Online computer museums have large educational, cultural and aesthetic significance. They promote preservation of our historical and cultural heritage. Edward Proydakov's Virtual Computer Museum launched in 1997 (http://www.computer-museum.ru) is the most representative and professional one in Russia. The unique feature of this project is active participation of outstanding scientists, engineers and designers who created the first Soviet computers. Many of them led projects that later came into legend. Different aspects of this museum's activity are described in the article. Some other Russian projects and foreign virtual museums are also represented.

Keywords: Computer, history, virtual museum.

1 Introduction

As known, virtual museum is "a collection of digitally recorded images, sound files, text documents, and other data of historical, scientific, or cultural interest that are accessed through electronic media. A virtual museum does not house actual objects and therefore lacks the permanence and unique qualities of a museum in the institutional definition of the term" (Britannica Online). These collections take full advantage of the easy access, loose structure, hyperlinking capacity, interactivity, and multimedia capabilities of the World Wide Web.

Last years a huge quantity of virtual computer museums had appeared in Internet. In early summer 2006 Google returned above 21 million links by inquiry 'Virtual Computer Museum' - naturally, mostly in English, but in other languages as well. Unlike traditional museums, museum in Internet is accessible any time irrespective of time zones, it doesn't demand extensive areas and contains no bulky exhibits. It is especially important for Russia where there are no specialized museums of computer hardware, neither representative collections of mainframes or minicomputers samples. The largest in Russia collection of museum pieces on computer facilities development is available in Polytechnical Museum, Moscow (http://www.polymus.ru).

J. Impagliazzo and E. Proydakov (Eds.): SoRuCom 2006, IFIP AICT 357, pp. 201–207, 2011.

2 The Russian Virtual Computer Museum

Edward Proydakov's project fills up this gap. His Virtual Computer Museum (http://www.computer-museum.ru) is the most representative and professional one in Russia. This project was launched in 1997, when Proydakov's article 'Let's create a museum' was published in PC Week/Russian Edition #38, 1997. That time Edward was the Editor-in-Chief of PC Week/RE.

> *"Once a son of my employee, being the university student of second rate, has asked his mum, if there were any computers before Pentium processor. And I have understood, that the youth simply does not know history of computer facilities"*

explains Edward. Soon he got a real support from his colleagues and adherents. The unique feature of this project is active participation of outstanding scientists, engineers and designers who created the first Soviet computers. Now they form the Museum's Advisory board. Despite of very respectable age and health problems, they take part in monthly sessions, write and discuss memoirs and textbooks.

Started by Eduard Proydakov in 1997
Our Sponsors
[News] [About] [Articles] [Advisory board] [Contacts] [Russian version]
© Russian Virtual Computer Museum, 1997-2006.

Fig. 1. Virtual Computer Museum Home Page (2006) http://www.computer-museum.ru/english/

It is necessary to mention here Prof. Victor Przhiyalkovsky - a Chairman of Museum Council, USSR State premium laureate, Hero of Socialist Labour, the general constructor of ES computer series and other mainframes. Other members of the Advisory board are famous computer designers as well: Prof. Yaroslav

Khetagurov (special onboard computers for fleet and aircraft); Prof. Alexander Tomilin (BESM-6); Dr. Nikolai Brousentsov ('Setun' - the first and unique in the world computer with figurative symmetric system of numbers representation); Dr. Yuri Rogachov (M-1, M-4, M-10, M-13). Edward Proydakov is a Director of Museum.

Here are the words from Museum's welcome message: "We were driven by the intention to keep public the most unique documents portraying the 50 years of Russian computing, its success and withering. The materials we present, being kept secret for decades, are even now at times being obscured by those unwilling to remember the true history... The members of our Advisory board are well known for their achievements in computer science and electronics, many of them led projects that later came into legend. Our collection includes unique materials on the development of military, special-purpose and universal computers, systems and application software".

Fig. 2. A session of Advisory board. From left to right: Yuri Rogachov, Victor Przhiyalkovsky, Alexander Tomilin.

The memoirs and scientific biographies of 'pioneers', materials on histories of domestic computer facilities, specifications of many computers are placed on the museum site. This information is supplemented with English-Russian computer dictionary (by E.Proydakov and L.Teplitsky), one of the most authoritative in the country (11000+ terms). Other popular sections of a virtual museum are 'Calendar of events', 'Hall of Fame', 'Documents and publications', 'Calculations in pre-computer age', 'History of domestic computer facilities', 'History of computer facilities abroad', 'History of software development', 'Technology', 'Computer games', 'Books and press'.

The activity of Museum is not limited to supporting a site, it often 'goes to offline'. Joint actions with the Polytechnical Museum took place. Members of Advisory board took part in exhibitions Softool at the All-Russia Exhibition Centre. The textbook on history of domestic computer facilities was prepared for edition.

In 2006 representative international conference was held in Petrozavodsk under IFIP. Virtual Computer Museum was one of the basic organizers of conference, and the Museum Council formed its Program Committee.

3 Other Russian Projects

Academician Ershov's Archive for the History of Computing[1] is a project of the A.P. Ershov Institute of Informatics Systems of Siberian Branch of Russian Academy of Science (Novosibirsk). Andrei Ershov (1931-1988) has left a unique archive of approximately 500 thick office folders storing documents which reflect his scientific career and simultaneously the history of computer science in Russia and worldwide. After this project was started in 2000, a lot of materials were digitalized, and now archive contains about 35000 documents, 120000+ images and about 6000 persons described. It is a colossal collection of digitized documents connected with scientific, administrative, pedagogical work and daily activity of an outstanding Soviet researcher.

Some organizations - computer developers have created 'corporative museums', web-pages with descriptions the history of their development. So, the Lebedev Institute of Precise Mechanics and Computer Engineering describes BECM, M-20, M-220, BECM-4, BECM-6, Elborus and other computers created in IPMCE. Russian version[2] is more detailed and illustrated.

The Virtual Museum of informatics[3] was created and supported by teachers and pupils of a school #444, one of the best physical and mathematical secondary schools in Moscow. Part of information is borrowed in E. Proydakov's museum. Most of other Russian virtual computer museums have poor original information. Some of them were launched by grant support and now have no renewal. The Museum of Soviet Computers history was created by Associate Professor S.Tarkhov from Ufa with grant support from Soros Foundation in 1999. It contains descriptions of computers, their characteristics, pictures and photos. However this project is not developed now. The 'Unofficial museum of computers in USSR'[4] (Denis Yakimov from Yaroslavl) has the similar destiny. A project 'Virtual museum of computer technologies' was initially a mean for distance learning in the chair 'Projecting and Technology of Computing and Telecommunication Systems' of Bauman State Technical University. Its page[5] contained many computer descriptions and historical articles from special magazines, but it is not supported now.

Sergey Frolov's original project on Soviet Calculators History started in 1998. Now author has 145 models of Soviet calculators in his collection. 50 more models form a list of items that the author continues to search for. His sites are rich of good pictures and detailed descriptions.

---

[1] See http://ershov.iis.nsk.su/english
[2] See http://www.ipmce.ru/about/history
[3] See http://schools.keldysh.ru/sch444/museum
[4] See http://old.h1.ru
[5] See http://museum.iu4.bmstu.ru/project.shtml

4   Russian Computers Abroad

One more virtual 'Museum of Soviet Calculators' was created by Andrew Davie, a collector of hand-held computing devices such as slide rules and electronic calculators. His project isn't so representative as previous one, but the author is excused because he lives too far from Russia, on Tasmania island. By the way, Sergei Frolov has provided most of the images on this site. Its address is: http://www.taswegian.com/MOSCOW/soviet.html, where Moscow means Museum of Soviet Calculators on the Web.

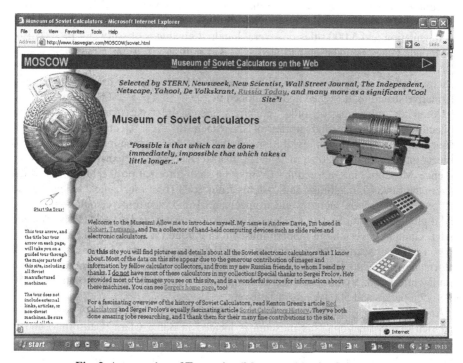

Fig. 3. A screenshot of Tasmanian 'Museum of Soviet Calculators'

Former Soviet citizens often put into Web their reminiscences about hardware and software they dealt with. For example, Leonid A. Broukhis (now citizen of UK) presents his 'BESM-6 Nostalgia Page' (1997-2005). This page is supposed to hold all sorts of information about great Soviet mainframe computer BESM-6. Address: http://www.mailcom.com/besm6.

The information on personal computer 'Agat'[6], a Soviet-made analog of 'Apple-2' can be found among 950 computer descriptions of 'Old-Computers.com', a project of Thierry Schembri and Olivier Boisseau. These two French men met in a pub in London in 1973. 25 years later they decided to launch this project (now hosted by NYI - New York Internet). Computer descriptions are available by name, by company or by year.

---

[6]   See http://www.old-computers.com/museum/computer.asp?st=1&c=509

5 Computer Museums in Different Countries

Foreign virtual museums (abroad) could be divided into two main groups: a) network representations of real (off-line) museums; b) 'pure' virtual collections exclusively accessible via the Internet. The first group is headed by Computer History Museum[7], the world's largest and most significant history museum for preserving and presenting the computing revolution and its impact on the human experience. Established in 1996, the Computer History Museum is a public benefit organization dedicated to the preservation and celebration of computing history. It is home to one of the largest collections of computing artifacts in the world, a collection comprising over 13,000 objects, 20,000 images, 5,000 moving images, 4,000 linear feet of cataloged documentation and 5,000 titles or several hundred gigabytes of software. In 2005 Bill & Melinda Gates Foundation had pledged a $15 million gift to the Computer History Museum. This museum is located on 7.5 acres of land in Mountain View, the heart of Silicon Valley; its new home is a two-story, 119,000 square foot architecturally distinctive structure, designed and developed in 1994.

Computer museums in other countries are mainly parts of more common collections of artifacts and stories of the Scientific Age. Deutsches Museum in München[8] has a computer exhibition on the 3rd floor which presents some 700 objects on an area of 1020 sq.m. The site has German and English versions. London Science Museum[9] has a section 'Computing and Information Technology' which covers the devices, machines and systems from 1623 (first mechanical calculator) to the present. Its scope includes electromechanical and electronic calculation, analog and digital computation, data management and processing, and cryptography. Science and Technique, a very large modern museum in Parque La Vilette (Paris, France) has now a special section on informatics on the second floor – 'Musee d'Histoire Informatique'[10]. Computermuseum in Netherlands[11] is open to the public every third Sunday afternoon. Visitors can see 25 mainframe computers, 10 minicomputers, 250 microcomputers and 500+ kinds of Software.

Among corporate museums we'll mention the Intel Museum[12], very professional and rich of information. This museum is located in Santa Clara (California) and is open to the public 6 days a week without any admission price. The Intel Museum collects, preserves and exhibits Intel corporate history for the purpose of increasing employee, customer and public awareness of Intel innovations, technologies and branding in an interactive and educational manner.

One more Project, the History of Computing Project[13] is incorporated as a nonprofit organization: History of Computing Foundation. It began in 1986 when somebody asked the original author (Cornelis Robat) during a course he gave: 'Where do computers come from and who started with it'. As we remember, the similar

---

[7] See http://www.computerhistory.org
[8] See http://www.deutsches-museum.de
[9] See http://www.sciencemuseum.org.uk
[10] See http://mo5.com
[11] See http://www.computermuseum.nl
[12] See http://www.intel.com/museum
[13] See http://www.thocp.net

question gave birth to Edward Proydakov's museum. This site contains a large list of hardware and its inventors; rich history of software and computer languages, a collection of historical papers by Pascal, Descartes, Turing, Shannon, Dijkstra and other ancient and modern authors. There are hundreds names in 'Biographies Pioneers of computing'. It also contains a good guide on other museums and even a collection of postage stamps with portraits of Babbage, Bernoulli, Ramanujan, Chebyshev etc.

Founded in 1995, the Virtual Museum of Computing includes an eclectic collection of web-links connected with the history of computing and online computer-based exhibits available both locally and around the world.

The Italian Computer Museum by Massimiliano Fabrizi contains mostly descriptions and characteristics of PCs from 1974 until the late 1980s as well as Polish site Historia Komputera.

The exposition of Ukrainian 'European Virtual Computer Museum'[14] mainly follows L. Malinovsky's books *The History of Computers in Faces, Essays on Computer Science and Technology in Ukraine* and *The Known and the Unknown in the History of Computer Technology in Ukraine*. Some sections such as The Computer History Abroad are fulfilled in cooperation with E. Proydakov's museum. We shall finish this list of the exotic countries with a mention about Mexican PC Museo and the Latvian Core Memory Museum[15].

6 Museums of the Internet

Internet became now not only environment and mean of information dissemination, but also an object of museum collecting. Let's specify some sources in Russia. Kurchatov Institute for Nuclear Research and Foundation for Internet Development recently introduced their project Museum of Internet History, which simulates a structure of real museum with exhibition halls (not all from them contain information now). Newest history of internet is represented by a research project called The History of the Internet in Russia[16], which purpose is gathering and analyzing information on development the Internet-technologies in Russia. It presents a chronology since 1990. The old versions of popular Russian and foreign sites are represented in 'Internet-Museum'[17], where you can go back to the past and imagine the internet as it was in the beginning.

Virtual computer museums promote increase of a general educational and cultural level and computer literacy; to preservation of our historical and cultural heritage.

---

[14] See http://www.icfcst.kiev.ua/museum/museum.html
[15] See http://www.thecorememory.com
[16] See http://www.nethistory.ru
[17] See http://museum.uka.ru

Computer Development in the Socialist Countries: Members of the Council for Mutual Economic Assistance (CMEA)

A.Y. Nitusov

Köln/Cologne, Germany; Moscow, Russia
nitussov@hotmail.com

Abstract. Achievements of the East European Socialist countries in computing -although considerable- remained little known in the West until recently. Retarded by devastations of war, economic weakness and very different levels of national science, computing ranged 'from little to nothing' in the 1950-s. However, full-scale collective cooperation with the USSR based on principles of equal rights and mutual assistance was aimed at increasing of common creative power. Centralised planning and ability to concentrate efficiently national resources on priority issues, state support for science and progressive educational system accessible for everybody played decisive role. The progress was impressive. Some (GDR) reached world's level in science and engineering such as some (in Hungary) – advanced computer education, programming and efficient usage and some (in Bulgaria, Cuba) starting "from zero point" turned into reputable manufacturers. In 1970-1990, 300,000 people as the united team of eight countries jointly designed and produced advanced family of compatible computers ES. Given general review also displays some important technical and organisational details.

Keywords: Computer development, East European countries, computer family ES, free education, cooperation.

1 Introduction

Little was written abroad computer development in the East European[1] socialist countries – partners of the USSR, during relatively long period, although some of their achievements were of considerable interest.

The lack of foreign attention was primarily caused by natural desire to display first the own pioneer discoveries and most important inventions. The results of the East-European (outside the USSR) computer research appeared notably later than in the "great powers", what could be another reason of "the silence". Besides, long time in the beginning eastern countries did not consider themselves as generators of pioneering projects or principal technical achievements. Additionally the cold war

---

[1] The name 'East Europe' to be often meet in this report means only the Socialist countries, such as GDR, Hungary, Czechoslovakia, etc., not including the USSR.

J. Impagliazzo and E. Proydakov (Eds.): SoRuCom 2006, IFIP AICT 357, pp. 208–219, 2011.
© IFIP International Federation for Information Processing 2011

confrontation and propaganda hampered obtaining and publishing appropriate authentic information. No wonder that only few - usually fragmental - materials sporadically became accessible for the Western reader.

As for more popular foreign publications or mass media, they were mainly influenced by the large computer manufacturers, interested first of all in their own publicity.

Otherwise, what could be the explanation of the fact that many educated young people in Western Europe were pretty sure that computer history was concentrated around IBM, the Apple Macintosh, or "Microsoft". It was often a surprise (e.g. for the West German students) that the world's first computer was devised in Germany by Konrad Zuse, or, say, that the Netherlands also created interesting projects?

Even situation in the USSR - one of the world's two biggest computer producers, equally remained almost "terra incognita" until the last decade. This time anti-Soviet cold-war propaganda coincided with atmosphere of exaggerated secrecy in the "Eastern block" itself. No wander that the first comprehensive publication on this subject appeared in the West only at the threshold of the 21st century.

Detailed demonstration of the computer history of all socialist countries can not be given in a single concise report. The present paper is not an attempt "to show everything". It is just a review written to help the reader to get a better understanding of the general course of computing history.

Although the East-European computing is not rich in scientific "sensations", it is interesting to observe the whole process of development in various aspects. Their economic potential was mainly incomparably lower than that of the "powerful West" or the USSR and their science and technologies were sometimes at a "beginner's" level. According to the initial conditions East European countries could be divided into following, approximate, groups.

The USSR possessed powerful economy, science, and performed progressive computer production.

The German Democratic Republic (GDR), Hungary, Poland, and Czechoslovakia had general scientific power and experience but little or no computing. Their economies were weakened by the war.

Bulgaria, Romania, Cuba, Mongolia, and Vietnam had neither proper science nor technologies applicable even for producing the simplest electronic devices.

Most of historical reviews and essays on computers and computations typically focus themselves on *scientific* and *technical aspects*. As their authors are mainly computer specialists or even the inventors themselves, they can provide the most precise, authentic and complete information and comprehensive explanations.

However, the role of general *social factors* in computer development also should not be neglected. Computer research and engineering is most complex subject, involving broad spectrum of scientific and other issues, including even humanities, when it deals with programming. Although influence of the social factors on scientific discoveries or technical inventions is very disputable and abstract issue, in the problems of (computer) proper application or of research and production organisation it can already be traced, while their impact -not necessarily direct one- on the general course of progress is often decisive. That can be clearly displayed by the example of East European (CMEA) countries, with their "alternative" social structures.

All of them managed to concentrate efforts and achieve impressive results within historically short time. Instead of training people to use imported computers (more typical for developing countries) they established their own research and production and developed efficient implementation. Besides economic benefits, that considerably stimulated development of national science.

No need to say that scientific and economic assistance of the USSR was very important factor. However, one cannot speak about any "pressure", especially in scientific or humanitarian area. Relations between the CMEA partners were based on the principles of equal rights, mutual respect and assistance with general motto, "common strength for the common success".

None was interested in turning others into raw material suppliers, assembling work-shop with cheap work-power or a trivial consumer market.

Of course, forming cooperation on practice was not easy. Different dynamics such as bureaucratic inertia and political problems caused complications. However, results speak for themselves. The first steps were made in the 1960-s. In 1972 all CMEA members adopted agreement on common development and production of the compatible computer series ES. Joint international coordinating commission was appointed. United team of more then 300,000 scientists and specialists from all countries were running their own ES program in 1972-1991. They successfully produced hard- and software and solved, "their own problems with their own resources".

At the same time the more powerful Western Europe, was weakening by inner competition. Many firms could not resist the pressure from the stronger IBM "partners" and, by the end of the 1970-s, had to close their own computer production.

Science and education of the CMEA countries became the fields to demonstrate advantages of the state support and centralised control/planning system. They enabled quick concentration of national resources and efforts on the programs of state priority, or important for perspective, even if they could not always immediately "pay back" in terms of investments.

Equally important were the absence of resource-wasting competition and government (or national) property in industry, that enabled high-tech equipment production on much lower investments. Private manufacturers would charge maximal prices for government orders in such case.

Accessibility of the highest level education and culture for everybody was one of basic programs, or constitutional principles, everywhere (in the East Europe). Propaganda and public opinion on education and culture were extremely high. They were considered as the criteria of social progress; success of an individual in education or scientific work was publicly estimated as his or her personal contribution to the common prosperity. Owing to cooperation, each country could regularly send its young people to the universities of the other partners. Thousands of foreign students permanently studied at universities of the USSR[2] or the GDR. Owing to that, an enormous mass of highly qualified specialists ("intellectual reserve") grew within one generation, not only in advanced Czechoslovakia and GDR but also in Bulgaria, Cuba, etc. That became their "Key to the door of computer society".

---

[2] As an example, in 1974 alone the USSR accepted more then 20,000 students from other socialist countries.

Important was also the fact that success of the national science had positive, stimulating, influence on the moral atmosphere in the whole society, what in its turn increased integral creative potential of the country.

2 German Democratic Republic (GDR)

Previously most of the East Germany belonged rather to agricultural, not industrial area. It was devastated by the war. However, German technological traditions, high-quality manufacturing, advanced engineering and rich scientific and educational heritage survived and formed the basis for rebuilding further progressive development of the country. Besides, one should not forget that in pre-war time, Germany already won the first place on the list of computer pioneers. It was the young engineer Konrad Zuse, who devised the first programmable computing machines Z-1 and Z-2 (with programs stored in memory) by the end of the 1930s, and, between 1942 and 1945, he created a system of commands named 'Plankalkul', recognised by specialists as the first high-level programming language. German researches on electronics, especially on semiconductors, were also in progress as early as in the 1920s and the 1930s.

Post war restoration of the Academy of Sciences (former Academy of Sciences Prussia), with its more than a 250 year-long tradition, as the GDR Academy of Sciences and its further development was one more important factor. Its work in the GDR was very efficient. By 1991 it consisted of fifty-eight research institutes with the total number of about 22,000 collaborators. After "integration" into West Germany (which had no academy after the war), the academy of GDR was dissolved. However, owing to the efforts of its members, it was partly revived as a scientific society.

Creation of new state, revival of destructed economy, and development of an electronic industry, practically from the zero point, needed time. The first practical steps to creation of the GDR electronic computers (outside theoretical and laboratory experiments) were taken in 1963, according to the new state program passed by the 6th Congress of the ruling Socialist United Party of Germany (SUPG) (15-21.01.1963). It read,

> "Herewith the Congress announces the beginning of new reformation process, focused primarily on the state economy, problems of the young generation and culture".

The "Development of the electronic data processing systems" was mentioned as a special direction of national efforts.

The result of that, the electronic (transistor) computer called 'Robotron R-300' (speed 5000 ops, RAM capacity 40 Kbit) was produced in Karl-Marx-Stadt (Chemnitz) in 1965. The R-300 was jointly created by such organisations such as the institute of electronics in Dresden, Karl Zeiss factories in Jena, enterprises ORWO, the office machinery factory in Sommerda, the Dresden institute of data processing. In 1965, twenty-two participants formed a cooperative enterprise "Robotron-300" (since 1969 -"VEB[3] Kombinat Robotron"). It was headed by "VEB Rafena" of Radeberg.

---

[3] VEB – Volkseigener Betrieb- (lit.) "enterprise owned by people" – a socialist form of collective ownership.

Within 1968 - 1971, about 350 "Robotron-300" machines were manufactured. "Robotron" was very powerful enterprise and it was established with the purpose of bringing GDR's lagging behind the Western competitors to minimum.

By the end of the 1960s, the GDR was the most active to promote project ES/ESER[4] because (besides other reasons) by that time it already started its own development of the IBM compatible computers on basis of the R-40.

The next computer, the R-21, appeared in 1970, but soon it was replaced by the first large German machine of the ES/ESER series. VEB "Robotron" was in charge of scientific researches in all computing related branches and also of numerous joint projects with the USSR for the ES/ESER computers. Serial production of ES computers began in 1973 and, by the end of the 1970-s, their total number exceeded 50% of all German computers. Within the years between 1972 and 1975, the number of control computers for production processes grew more than ten fold.

"Robotron" produced processor units, peripheral devices, and data transmission devices, managed the sale of office machinery factory "Zentronik" production and was also in charge of the data processing systems, their sales, implementation, and maintenance.

In the 1970s, new enterprise "Gruna" was founded in Dresden, it was responsible for production of central processing units for the R-21. "Robotron" was constantly growing and it was always efficiently re-organised in correspondence with new needs.

Development of microelectronics and the following reduction of computers' sizes changed character of its manufacturing. The government passed a proposal (resolution on microelectronics of the SUPG central committee, from 1977) to decentralize management of the computing machinery applications. For that purpose, an enterprise on microelectronics was founded in Erfurt in 1978. In 1978, its eight-bit central processor unit U808 was taken by "Robotron" as a basic component for microcomputer ZE-1 and special system K1510, which was followed by the K1520 with central processor U880 (Zilog-80 was its prototype). The U880 processor was also implemented in office computer A-5100 and later, in the 1980-s, in a very popular PC "Robotron-1715", which was in commercial production until the end of the GDR itself.

In 1986, computers also appeared in retail trade (microcomputer set Z1013). Microcomputer systems K1510 and K1520 were especially popular in banks and railway terminals. By 1989, "Robotron", with its twenty-two enterprises-participants and almost 77,000 personnel, was the biggest computer manufacturer in the GDR.

The 10th Congress of SUPG defined the program on increasing efficiency of computer operation organization. According to it, the computers should be used during several working shifts a day. Operational efficiency of computers was also increased by extension of their application range. The first computers were used for registration, later in research, and still later in technical preparations for production. Level of the personnel qualification was also stably rising. According to the official data of 1974, economical efficiency of computers exceeded 1 Mio German Mark a year. In 1975 an average operation period of the computer at an enterprise equalled 15.6 hours a day and in some branches even 18 hours per day.

---

[4] ES (Russ.) Edinaya Seriya – Unified Series, ESER (Germ.) – Einheitlihes System Elektronisher Rechentechnik.

Within ES/ESER program GDR specialised its participation on development and production of central processors of medium power and memory devices on magnetic tape. The first computer ES-1040 was produced very actively and was exported to seven countries. In 1976, central processor ES-2640 was given a special award for its high quality. Owing to cooperation with the USSR, Karl Zeiss factories in Jena quickly established production of the magnetic tape storage devices. Before 1976, GDR sold 13,000 such units. The ES-2640 storage devices also received awards for their high quality.

The GDR took active care of solving organisational problems within CMEA and actively promoted strengthening of partnership between the ES/ESER program participants. It always insisted on introduction of common unified system of quality and reliability evaluation and intensifying software exchange between the partners.

Large computers R-55 and R-57, produced by the GDR belonged to economically important machines; large numbers of them were exported to the USSR in the 1980-s. Equally, much attention was paid to the software production. In 1987, a part of "Robotron" in Dresden became the leading centre of the software development for the whole GDR.

3 Czechoslovakia

In the beginning most of the work on electronic computing machinery was closely connected with the name of Antonin Svoboda, a pioneer of control and analog computers. Owing to rich experience that he received while working in Paris, in the USA, and also at the Cambridge radiological laboratory, Svoboda was able to make a big contribution to a new science. Thus, the institute of computation machinery was established on his initiative at the CzSSR Academy of Sciences. A number of (later) prominent scientists began their career in that organisation. Also the "heritage" of Prague factories for precession machinery played important role in the creation of national computer manufacturing basis. Thus, one of them – the firm "Aritma" – produced punch-card machines. In 1951 a national enterprise for office equipment was established. It was responsible for service and distribution of all Czechoslovakian computer hardware.

The first –relay- computer SAPO (*SAmochinny POchytach*) with good parameters for its time was created within the period 1949–1957, at the laboratory of mathematical machines, which later grew into the Scientific Research Institute of Mathematical Machines (VUMS). After that, VUMS developed computers MSP-2, EPOS, and ZPA-600 and numerous peripheral devices. The "National Eliot 803A", 803B", and 503" computers were purchased in England for research purposes as well as some models of the Soviet Minsk and the Polish Odra machines.

In the period from 1949 to 1974, the total production of "Aritma" consisted of approximately 18,000 punch-card micro-calculators "Aritma" (Aritma-100, Aritma-1010 and Aritma-101), about 220 computers, also 1000 analog computers MEDA, and other 160 special single-program analog machines.

Quite naturally, East European countries revitalised their traditional international connections. Thus the hybrid systems HRS - Robotron 4241, HRS 7200 and HRS 7000 were jointly produced with enterprises of the GDR. Czechoslovakian national enterprise Tesla began production of the second generation computers on the licence of French company Bull. Parallel with "Aritma" computers were produced by the "Zbrojevka" factories in Brno (typewriters and mechanical calculators), factories for instrumentation and automatic systems in Koshize, and Tesla in Prague. Nevertheless, by the end of the 1960-s the proper level of stable production, that could meet national needs, was not yet reached.

The situation changed for better only in 1969, when the production was re-organized. Thirteen leading manufacturers began to be integrated into a joint system. Besides Tesla, "Zbrojevka" factories and others, also the sales organisations, such as "Kancelarske stroje" in Czech Republic and "Datasystem" in Slovakia, provided some important services. All that integrated structure was headed by the Prague enterprise of automation BT-Zavt aimed at international cooperation.

Zavt took part in the joint production of the ES production and later in the series SM (ES-1021, ES-1025, SM-1, SM-2, SM-3-20, SM-4-20) (family of smaller computers). It also manufactured microcomputers and microcomputer systems, analog- and hybrid- computers, such as MEDA-41TC, MEDA-42TA, MEDA-43HA, ADT-3000 and HRA-4241.

Some participants produced peripheral devices for the ES and the SM, others data processing systems and devices and also office machinery (e.g. typewriters Consul). Zavt was responsible for design and production of popular automatic control systems, instrumentation and scientific equipment for laboratories.

Scientific researches were performed by VUMS, by the Institute of mathematical means (VUAP), and by the institute of computing machinery applications UAVT in Prague, by the institute of computing (VUVT) in Zhilinka and by a research centre in Bratislava.

Participation in the ES program was based on the following doctrine: Czechoslovakia concentrated on production of such ES computers and equipment that could first of all satisfy its own needs. It exported its peripheral devices that won good reputation, to the other partner countries and obtained money was mainly invested into further development of its own computer industry.

The ES-1021 computers (or R20A) of the Prague VUMS were designed for solving scientific, engineering, and economic problems of industrial enterprises of medium capacity. However, they could also perform functions of satellite computers for large machines. They were produced by factory ZPA in Zakoviza, famous for the minicomputer MSR-2 and medium machines ZPA-600 and ZPA-601. Aritma of Vokovize produced perforators and punch-card reading devices.

In 1974 Czechoslovakia took new series of urgent measures. According to new (fifth) five-year economic development plan, 340 computers would be manufactured. Three hundred of them would be ES computers. The (sixth) five-year plan suggested further increasing of production to 800 - 1000 pieces.

CSSR became one of the leading manufacturers of the peripheral devices, in particular magnetic tape reading units, for the world market. All export operations were performed by the national enterprise KOVO and delivery, assembly, and testing were made by *Kanceljarski Stroje* (KSNP).

4 Hungary

Already in the pre-war time, Hungary became a motherland of scientists famous in computer world. However, most of their work was done abroad. Thus Laszlo Kozma (1902-1983) was working upon a relay computer at the Bell Telephone Laboratories in Antwerp, Belgium, in the 1930-s. He devised it by basically using telephone components.

Janosch (John) von Neumann (1903-1957) was born in Budapest, where he received education and also his doctoral degree in mathematics. Between 1926 and 1930, von Neumann held position of *privatdozent* in Berlin. In 1930 he married Mariette Kovesi in Budapest and soon, upon the invitation of professor Osvald Weblen (USA), moved with her to the Princeton University, where he spent the rest of his life. World famous are his theoretical principles of computer architecture.

The post-war computer researches started in the universities of Budapest and Zeged. In 1957, the Hungarian Academy of Sciences received from the USSR Academy of Sciences a complete set of project documents on digital electronic computer M-3, created by Moscow academician I.S. Bruk and his team. Assembled in Budapest M-3 became the first Hungarian electronic computer. Nevertheless, during the following years, computers were mainly imported. Purchasing of licences was also practised in later years. In 1968 the French company CII (which "inherited" famous BULL) handed Hungarian specialists the rights for computer CII-10010. In France itself that model was developed into a series of civil and military computers.

In the beginning of the 1960-s, Hungary had only ten computers and by 1967 it had eighteen. Participation in the Council for Mutual Economic Assistance (CMEA) noticeably intensified that process. In 1970 there were 80 machines; in 1972 Hungary had 120 machines; in 1975 the number reached 400 and by 1980 more then 700 machines. The "new period" began in 1969, with the ES project. After a short time, Hungary launched production of the ES computers and later minicomputers of another series - SM. In parallel with it, Hungary maintained relations with Western computer manufacturers such as Siemens, CDC, and UNIVAC. Various forms of development were well combined and successfully supported each other. The alphanumeric display Videoton VT-340 / ES-7184 and the minicomputer VT-10108 are famous examples of that support. The agreement with the French CII of 1971 focused on production of the ES compatible minicomputer of the third generation.

Production of the ES computers started in 1972. First those were ES-1010, then ES-1010-M, ES-1011 and ES-1012. Their export began in 1973-1974. Besides the computers themselves numerous printers, intellectual terminals and punch-card input devices were also manufactured. Computers ES-1015 and SM-53 appeared on market already by 1979.

According to the governmental information, about a hundred leading Hungarian enterprises possessed their own computers by 1979 and 1500 more regularly rented working time at the computer centres of collective usage. Governmental reports emphasised high efficiency of the computer implementation in national economy and science.

Development of control and information systems, as well as the coordination of computing machinery production and implementation measures, became the processes of national importance and dimensions. Timely getting special, highly professional organisations in charge of those processes brought a credit to their initiators. In 1976, an independent self-financing scientific-research institute for applied computation systems (SZAMKI) was established with a staff of 500 collaborators.

SZAMKI was a competent organisation that maintained scientific relations and performed both local and foreign projects. SZAMKI performed projects on the ES operation systems, established new computer centres, designed computer networks for informational and translation systems, ran a large project in the field of qualification upgrading, published computer books and journals and was very active in cooperation with other CMEA countries. In general, this institute played very important role in Hungarian computer development.

Hungary always had very good contacts with the GDR, which were supported by the bilateral agreement on "Development and joint production of the machinery for remote data processing, data banks creation, design and production of processor units, microcomputer modules and line printers" adopted in 1972. Relations with the USSR and CSSR were also successful.

According to the new development program for electronic industry of 1981, seven enterprises became suppliers of its production. The Hungarian Academy of Sciences controlled the research work.

Development of the programming products became another strong point of Hungarian computing. In 1969 the national educational centre on computing was established. There, young scientists established a school of programming in cooperation with the University of Budapest. With the support of the UNO, it quickly grew into a very popular "International Centre of Computer Education" (Hungarian - SZAMOK). It was officially opened in 1973. In the same year the school received more than 8000 applications. Although SZAMOK had only one computer IBM-360/145 and one PDP-11/70 - a machine with multiple accesses in time shearing mode, it managed to achieve notable results, both in general computer education and in system programming, corresponding the level of the best world standards of that time. Its education was accessible not only for Hungarian, but also for foreign students. In addition to Hungarian, all courses were available in German, Russian and English languages. During the period from 1969 to 1978, SZAMOK gave education to more then 52,000 specialists. That activity created perfect conditions for establishing new computer centres and programming firms.

Most important was that, those efforts turned Hungarian computing into independent science. One more independent organisation also made sufficient contribution. It was the scientific "Janosch von Neumann Society", founded in 1975 that dealt with programming, algorithms, system organisation, etc. Its basic task consisted in "support of the centralised state program on computing development". The society published two periodic journals "Computation machinery" (Szamitatstehnika) and "Informatics and Electronics" (Informacio es Elektronika), which won special support.

Popular TV courses of the BASIC programming language turned to be one more successful program. More than 3000 of its 6000 applicants successfully passed graduation exams. The government also supported numerous computer clubs.

In the 1990-s some Western specialists, programmers, and simply those interested in "computer matters" in the East, wrote (in internet) that they could only dream about such public atmosphere so good for children education and general development. Rapidly spreading "computerisation" still supported successful programming branch, sufficient part of whose products had comparable or even equal quality level with software found in Soviet or Western markets. Hungarian applied programs on general, special, and also unusual subjects were regularly demonstrated at the international fairs in Leipzig and Hanover.

5 Poland

Same as Germany and Hungary, Poland had a solid scientific potential and traditions both in mathematics and calculation devices. However, its economy was greatly undermined or rather almost completely destroyed by the war. Despite sufficient support from the USSR (that also suffered from heavy losses) any serious production of electronic devices seemed to be unthinkable that time. Nevertheless, development of Polish computing began as early as in 1948 at the Institute of Mathematics in Warsaw, where the group of mathematical devices (*Grupa Aparatow Matematycznyh - GAM*) headed by Henryk Granevskiy designed analog computers for solving differential equations. Their computer (*Analizator Rownan Rozniczkowyh - ARR*) with 400 electron valves was completed in 1954. Appearance of the first Polish computer sufficiently facilitated work of mathematicians.

Generally speaking, the "Polish way to computer" followed more or less the same pattern as in the other Eastern European countries. In the beginning very poor quality of electron valves became serious obstacle. First the problem was solved by traditional imports. As a result of that, the XYZ digital computer appeared in 1958. Although its architecture repeated that of the IBM-701, design of its basic elements was taken from the Soviet BESM-1. An improved model of XYZ, named ZAM-2, was already a large machine with performance of 1000 ops (both floating- and fixed-point operations). Some of its units were even exported. Its Special System of Automatic Coding (*Systema Automatichnogo KOdowania - SAKO*) was the specific feature of the both computers. Efficient programming system, often called "Polish Fortran", was created in 1960.

Nevertheless, main period of computer development began with joining CMEA and the ES program. Poland produced computers ES-1030 and, after a number of improvements, manufactured famous ES-1031 and ES-1032. It also took part in development of software for ES computers. Between 1971 and 1975, Poland began commercial production of fourteen various devices, five of them were implemented in various ES computers. Those were magnetic tape storage units (ES-5001), user node (ES-8514), programmable MPD (ES-8371), modem (ES-8013), and a collective system for data preparation on magnetic tape (ES-8013). In addition to ES computers, Poland also produced its own series of computers MERA.

6 Bulgaria

In the beginning of the 1950-s, Bulgaria did not nave any electronic industry and could not even produce simple components for radio. It was typical agricultural country. Nevertheless, the government launched national campaign for total education and took all necessary measures to facilitate it. Education of all levels became free and accessible to everyone. There were only few universities, so hundreds of Bulgarian students regularly studied at many universities and institutes of the USSR, the GDR, and the CSSR. Together with the students from, Poland, GDR, Hungary, Cuba and Viet-Nam numerous young Bulgarians were very common in almost every big technical Soviet university. Already in the 1960-s special attention was given to preparing electronic and computer specialists. The first Bulgarian computer centre was opened in 1961 and the first computer VITOSHA, was made in 1963. Electronic industry was gaining power. The first Bulgarian transistors were produced on Japanese and other licences. By the beginning of ES program, Bulgaria already possessed necessary scientific potential to be an equal (and very active) partner. Relations with the USSR traditionally were very close on all levels. Bulgaria participated in the joint production of the ES-1020, then the ES-1022, and then the ES-1035. National enterprise ISOT integrated fourteen other large manufacturing enterprises and a number of research centres. Within one generation, Bulgaria managed to grow to be an important partner in computer production. It became a recognised manufacturer of magnetic memory carriers (tapes and discs), sufficient amount of them was exported to the USSR. In the 1980-s, especially with the appearance of minicomputers and PCs. Memory devices and magnetic discs of Bulgarian ISOT became very popular among the Soviet computer users. In the 1980-s, Bulgaria also produced its popular PCs called PRAVEC.

7   Romania, Cuba, Mongolia, and Viet-Nam

Production and implementation of computers in these countries developed mainly in the same way. In the beginning their scientific and engineering basis was prepared by systematic sending of their students to study in more advanced countries (USSR, GDR, etc.), then followed development of their own researches and projects (at first, minor ones).

For an example, Cuba before the Revolution of 1961 possessed only two (imported) computers. In 1969, the Centre of Digital Researches (CID) and some other institutes were found and the first Cuban minicomputer CID-201-A was in operation in 1970. It was designed by Orlando Ramos, pioneer of Cuban computer engineering. Cuba took part in the ES program what helped it to create certain electronic industry. Efficient state programs on mass computer education (including school programs) were introduced including obligatory practicum for school-children who should perform various tasks at computer centres and work with computers. Clever educational policy and state measures on propagation of computer knowledge soon created "new generation" of people basically prepared for learning and working in computer-related fields. Although Cuban engineering did not reach the level of big machines, later it launched production of its own minicomputers and various digital devices of good quality. Especially popular was –and still is- its digital medical equipment that is actively exported.

Notably, nowadays Cuba remains the only country, which –to certain extent- is preserving "scientific spirit of the CMEA" and developing its engineering traditions.

Despite limited scientific and engineering achievements, experience of these countries in *computer implementation* was undoubtedly interesting. Besides technical projects, mathematical researches and software development, efficient organisational measures were undertaken on governmental level, which created favourable conditions for intensive "computerising" of these, formerly pure agricultural, lands.

In the beginning Romania made efforts to develop independent computer industry, but with only moderate success. It produced several models of minicomputers "Felix" on French licences, then joined the ES program but was not active in it.

November, 2009.

-     For more information on the topic, please address
  http://www.computer-museum.ru/
- =   Additional materials, comments and possible corrections will be gratefully accepted by the author: (nitussov@hotmail.com)

References

1. Trogemann, G., Nitussov, A., Ernst, W.: Computing in Russia. VIEWEG Wiesbaden (2001)
2. Rakovski, M.E. (ed.): Computer engineering in the socialist countries (periodical journ.). Statistics (1-8) (1977/1980)
3. Geschichte des VEB Kombinat Robotron. Zusammengestellt durch Frau Dr. Kretschmer
4. Malinovsky, B.N.: History of computers in personalities. Kiev KIT. A.S.K. Ltd. (1995)
5. Communism and Computing. Comm. of the ACM 35(11), 27–29, 112 (1992)
6. Russian Virtual Computer Museum (on-line) materials,
   http://www.computer-museum.ru

On the History of Computer Algebra at the Keldysh Institute of Applied Mathematics

G.B. Efimov, I.B. Tshenkov, and E.Yu. Zueva

Keldysh Institute of Applied Mathematics, RAS, Moscow
{Efimov,Ezueva}@keldysh.ru, Iesm2@mail.ru

Abstract. The authors consider the history of Computer Algebra (CA) development and applications at the famous Keldysh Institute of Applied Mathematics (Russia Academy of Science). At the Institute CA was used in various areas: Applied Celestial Mechanics, Mathematics, Robotics, Hydromechanics, and Methods of Calculation. Authors compared and classified the existing systems and their possibilities. An attention is paid to REFAL (REcursive Functions Algorithmic Language), developed in the Institute. The modern state of investigations is also concerned. The list of the main references is presented.

Keywords: Computer algebra, mechanics, mathematics, programming, history, KIAM, REFAL.

1 Introduction

The Keldysh Institute of Applied Mathematics of the Russian Academy of Science (KIAM RAS) was the first and one of greatest centers of computer science in the Soviet Union. M.V. Keldysh founded it in 1953 for solving difficult scientific problems of national importance. From the very beginning, it was the problem of modern physics, from mathematical and computational point of view. Later the Institute obtained the significant results in such scientific areas as space mechanics, physics, mechanics, cybernetics, programming and others. At the Institute, the experts in different areas such as mathematicians, physicists, mechanical engineers and computer scientists worked together, and their contacts with each other became the source of fruitful ideas. Such famous scientists as A.N. Tichonov, K.I. Babenko, I.M. Gelfand, T.M. Eneev, A.A. Liapunov, D.E. Okhotsimsky, A.A. Samarsky, Ya.B. Zeldovich, V.S. Yablonsky, K.A. Semendyaev, O.B. Lupanov, M.R. Shura-Bura, A.N. Miamlin, S.P. Kurdiumov, N.N. Chentsov were among them. Numerous difficult problems required various mathematical methods for their solution. These scientists were using the early computers and they were experiencing the early successes in this new field. These successes, in turn, gave all of them the enthusiasm to generate many interesting ideas.

Soviet programming, different directions of computer science, and modern methods of computations were born in this first period. It is interesting to mention that

J. Impagliazzo and E. Proydakov (Eds.): SoRuCom 2006, IFIP AICT 357, pp. 220–227, 2011.

our Institute was among the first users of new computers. We received Soviet-made computers 'M-20' with serial number 1 and 'BESM-6' with serial number 2. We also had the opportunity to use some others Soviet computers such as 'STRELA', 'BESM-3' and 'BESM-4', and various computers from the 'ES' series. When programming as a new kind of activity had appeared, it required new people. A number of professional programmers worked in the Institute; they created new languages, translators, systems, and packages. Note at that time, that applied scientists, physicists, and mechanical engineers stimulated this process and participated in it together with professional programmers. They also were the authors of software in the 1960s, 1970s, and later.

2 Applications in Computer Algebra

The idea of symbolic computations arose just after the appearance of computers. It started as a will to make computers able to interact with humans via an ordinary mathematical language and to facilitate the job of physicist or mechanical engineer. Thus, the history of computer algebra (CA) in the Soviet Union and Russia spans a half century. This history is presented in several reviews and conferences papers [1-17]. This list includes some works of authors of this paper devoted to the application of CA and CA systems in mechanics, dynamics of multi-bodies systems, theory of control [9, 15, 17] and to CA investigations at Keldysh Institute [16]. The review [17] contains detailed references to more then 500 original works.

It is interesting to remember the very first attempts to use symbolic, non-digital way of interaction with computer, at the age when most of scientists considered computer as a big and not very convenient calculator only. In their historical review, A.P. Ershov and M.R. Shura-Bura [2] mentioned only the first works of Leningrad's school. We will add the works of some other groups of authors made in Russia as early as the beginning of the 1960s.

In 1956 at the first National Conference devoted to computer science, A.A. Dorodnitcin formulated the task that we can consider as the beginning of symbol manipulation in the Soviet Union [18]. He proposed to build solutions in the form of two asymptotic power series, conjugated by a common numerical solution in the regular area. Thus, analytical approach was combined with numerical one. D.E Okhotsimsky from the Keldysh Institute applied this approach to a space dynamics problem [19]. He also built two asymptotic series conjugated by a common numerical part. Automatic building of asymptotic construction on one of the first Soviet computers, the STRELA, was innovation per se.

A.A. Stogny [20], the follower of V.M. Glushkov in Kiev, resolved the task formulated by Dorodnitcin. He proposed an algorithm to obtain the polynomial solution for a differential equation. It was probably the first result of the famous Kiev Computer Algebra School. At the same time, Nobel laureate L.V. Kantorovich investigated some types of symbolic notation for computer algebra in the Leningrad's branch of the Steklov Mathematical Institute. His follower, T.N. Petrova, created the program "Polynomial Prorab", used later for the tasks of theory of elasticity and other applications [21]. N.N. Yanenko, who started his scientific way at KIAM, investigated Cartan's methods of the analysis of compatibility of systems of differential equations in partial derivations and realized it on the STRELA computer [22]. In Leningrad, at the Institute of Theoretical Astronomy (ITA RAS), V.A. Brumberg and his colleagues developed solutions of some problems in celestial mechanics in series form [23]. Later, they elaborated and expanded this approach. The group of L.V. Kantorovich in Leningrad and V.K. Kabulov in Tashkent elaborated a similar approach to the tasks of the theory of elasticity [24].

The Keldysh Institute was among the first developers of symbol manipulation and began the first attempts to use them in the area of applied celestial mechanics. Z.P. Vlasova and I.B. Zadyhaylo implemented the first manipulations with trigonometric and power series on Soviet STRELA computer. In 1964, D.E. Okhotsimsky initiated work on a semi-analytical solution of a space dynamics problem for the low-thrust flight in a central field of gravitation. G.B. Efimov realized this approach for a simple Poisson series in 1970. Since 1970, A.P. Markeev used CA for normalization of Hamilton Systems and periodic solution stability analysis. His follower, A.G. Sokolsky, made the next steps in this direction. This work continued by the same group at Moscow Aviation Institute (MAI) and later by A.G. Sokolsky in ITA RAS in Leningrad. At MAI, they developed computer algebra applications to education as well. V.A. Saryshev and S.A. Gutnik used CA for the problem of satellite equilibrium stability in 1984.

Since 1963, M.L. Lidov with his group implemented numerous experiments concerned CA application to satellite dynamics problems. It was impossible to use existing general-purpose CA systems because they were very primitive at that time to solve such complicated problems. He proposed a method to unite both analytical and numerical approaches. For elliptic orbits and perturbations of different sorts, he used the analytical approach to construct the Hamilton perturbing function. Then, he calculated the right-hand parts of the perturbed motion equations in numerical form in every step of the integration. He implemented this calculation, using a Hamilton function derivation. This approach provided a high accuracy method for motion calculation and it allowed the avoidance of labor-consuming analytical calculations. As a result, very sophisticated special algorithms and programs were elaborated and used by M.A. Vashkoviak, A.A. Soloviev, Yu.F. Gordeeva, and other followers of M.L. Lidov. Unfortunately, these investigations did not stimulate the development of CA systems for common usage.

Let us mention some works in the area of fluid and gas dynamics. I.B. Tshenkov and Ya.M. Kajdan resolved several hydrodynamics problems with the aid of a REFAL-based CA program. M.Yu. Shashkov and L.N. Platonova applied similar methods using REDUCE. In the area of dynamics of complicated multi-body systems (e.g. robotics, spacecraft), they used CA for deriving the equations of motion, stability investigation, automatic generation of program of numerical analysis, and other applications. They created the PAS and others systems used for deriving equations of motion and for solving some problems of the theory of control. Computer algebra applications to mechanical education were developed by D.Yu. Pogorelov, the follower of V.V. Beletsky.

3 REFAL and Its Use

The very interesting page in Russian cybernetics is the history of REFAL. In 1969, V.F. Tourchin created REFAL - an original computer language based on a new principle of programming. He used associative text processing based on recursive function theory without directly addressing control of the program [25]. From the beginning, computer algebra was among potential areas of REFAL applications. However, the first realization of REFAL was more "scientific" then practical because it was isolated and not compatible with "ordinary" software such as numerical packages, library support, and memory allocation. Additional efforts of many people required to make REFAL modifications usable in a practical way, in particular for computer algebra applications. S.A. Romanenko and A.V. Klimov from the group of V.S. Shtarkman created high-effective compiler from REFAL [26].

V.F. Tourchin and others developed the first REFAL program to solve some problems of nuclear physics in series. This program demonstrated some CA features. I.B. Tshenkov elaborated a general-purpose CA system named SANTRA and later modified it. Later I.B. Tshenkov and M.Yu. Shashkov made the special applied system, called DISPLAN, for processing non-standard difference schemes. M.L. Lidov and L.M. Bakuma were among the first REFAL users in the applied areas. They attempted to process Poisson series. H.C. Ibragimov and I.B. Tshenkov made attempts to apply REFAL in the group theory. Initially, the plan was to use REFAL as a "meta language". However, people began using it in a wide scope of tasks similar to CA, where text processing was necessary. V.A. Fisun, A.I. Khoroshilov, and others developed computer languages based on REFAL, such as SIMULA-1 and DYNAMO. A.N. Andrianov and K.H. Efimkin automated the calculation of difference schemes in the NORMA system.

The KIAM united some of the enthusiasts of REFAL and they worked closely with it. V.L. Topunov with his colleagues from Moscow State Pedagogical Institute used REFAL-based CA system in differential geometry. Together with V.P. Shapeev and the other followers of N.N.Yanenko in Novosibirsk, they used the method of H. Cartane and investigated the characteristics of difference schemes. L.V. Provorov at the CAHI (Professor N.E.Zhukovsky Central Aero-Hydrodynamic Institute) and at the Bauman High Technology University, O.M. Gorodeskiy (Grodno) and A.V. Korlyukov created a CA system for numerous application areas in engineering. They could derive the equations of the motion of multi-body systems automatically. Use of REFAL allowed presenting the equations in convenient form near to human presentation. L.F. Belous and I.R. Akselrod (in Kharkov) used REFAL for the integration of several different programming systems and numerical packages into a united system; included in this system is the well known REDUCE package and the domestic computer algebra package called SIRIUS.

At the Keldysh Institute, some leading scientists investigated the possibilities of increasing computational efficiency. They did it by creating specialized blocks/computers, united in the complex system. This was the way to process effectively various tasks, numerical as well as symbolic; indeed, REFAL was appropriate for them [27]. A.N. Miamlin, I.B. Zadyhaylo, and V.K.Smirnov were the leaders of the project [28]. L.K. Eisymont analyzed REFAL efficiency for both software and hardware realization, and in particular from the CA application point of view. V.K. Smirnov with his group simulated the REFAL processor EC-2702 on the EC-2635 computer using microprogramming [29]. This processor was compatible with EC-series of computers and they used it for CA problems as well as for translators. K.I. Babenko, A.V. Zabrodin, and I.B. Zadyhaylo investigated the possibility of designing a highly efficient parallel computer for applied (aerodynamics) problems. Recently A.V. Zabrodin and others developers used this early result to create the Russian MVS-100 and MVS-1000 supercomputers; the MVS-1000 had an efficiency of about 10 power 12 operation per second [30]. The programmers successfully used REFAL-oriented software design techniques for creating supercomputer program environments.

4   A New Era

At the beginning of the 1980s, CA popularization and systems comparison became important, in particular for mechanical tasks that required computer experiment for their investigation and solution [8, 9, 17]. In 1982, scientists successively used CA for constructing difference schemes in the area of non-regularity. A.A. Samarsky, a leader in Soviet mathematical modeling, evaluated favorably the result of this work. With the aid of A.A. Samarsky, KIAM became one of the main organizers of the First National Conference on Computer Algebra Applications in Mechanics held in Gorky (now Nizhny Novgorod) in 1984 [7]. The participants of the Conference discussed the results of about twenty years of research, as well as plans and perspective directions of the subject. Later, a number of conferences and seminars on CA had taken place in Russia.

G.B. Efimov and M.V. Grosheva began to classify and to generalize the experience of the common work of mathematicians, mechanical engineers, and programmers. There were many reviews of CA systems and CA applications for mechanical problems [6, 9, 14 and 17]. The reviews provided users with tables of CA system features. They also provided a convenient tool for CA system comparison and selection for potential users — experts in applied areas. Such analysis was useful for CA developers as well. First, CA systems were usually specialized and elaborated for concrete task solution. For example, in the area of dynamics of multi-body systems they created many various CA systems, so the classification and comparison of their features became important. Later, well-known modern universal systems such as REDUCE and others appeared and people used them for the realization of algorithms. Currently, computing specialists use CA as a necessary standard tool in large complex programs, often without any special mention. Information about CA systems in mechanics was accumulated during the classification; this work has now become a valuable tool for applications and an important contribution to the history of computing [17].

Finally, we would mention some results obtained recently. A.D. Bruno and his group obtained new important results [31]. It concerned some algorithms of normalization in Hamilton systems and investigation of Newton polyhedrons. S.Yu. Sadov and V.P. Varin investigated the stability of motion for several problems of celestial mechanics [32]. A.S. Kuleshov [33] generalized the classical work by Chaplygin. Using CA he found some new cases of integrity and the first integrals in explicit form. A.W. Niukkanen and I.B. Tshenkov developed a system of transformations for hypergeometrical series [34]. Some authors used CA to generate the equations of motion of multi-bodies systems, and to simulate the dynamics of these systems. In the situation when it was impossible to use CA directly, some people developed other similar methods. For example, D.Yu. Pogorelov [35], I.R. Belousov and I.Yu. Balaban [36] developed such methods to simulate multi-bodies systems with a great number of degrees of freedom, such as robots and transport systems.

References

Almost all publications are in Russian, except noticed by "*". At some numbers the references to Russian "Reference Journal on Mathematics and Mechanics" was done as – "РЖ МаТ", "РЖ Мех".

[1] Calculus mathematics and computer facilities. All-Union seminar. Inst. of Low Temperature Physics AS UkrSSR (3), 152 (1972)

[2] Ershov, A.P., Shura-Bura, M.R.: Development of programming in USSR. Part 1. Initial development. Part 2. Step to second generation of languages and computers. Preprints No. 12, 13 of Computer Center of Siberian Branch AS URSS. Also: Cybernetika (6) 141–160 (1976)

[3] Computer analytical calculations and their application in theoretical Physics. In: Materials of International Conference, Dubna (1979); p. 187. JINR, Dubna (1980), p. 260 (1983), p. 420 (1985)

[4] Gerdt, V.P., Tarasov, O.V., Shirkov, D.V.: Analytical Calculations on ECM applied to Physics and Mathematics. Uspehi Phys. Nauk. 30(1), 113–147 (1980)

[5] *Miola, A.M., Pottosin, I.V.: A bibliography of soviet works in algebraic manipulations. SIGSAM Bull. 15(1), 5–7 (1981)

[6] Grosheva, M. V., Efimov, G.B., Brumberg, V.A., et all.: Computer analytical calculation systems (Applied Analytical Packages). Papers Seminar in Mechanics Institute of MSU, (1981); Informator. Keldysh Institute of Applied Mathematics AS USSR (1), 65 (1983)

[7] Systems for Analytical Manipulation in Mechanics. In: All-Union Conference, Gorky (1984); Abstracts. GSU, Gorky. p. 147 (1984)

[8] Applied Program Packages. Analytical manipulations, p. 156. Nauka, M. (1988)

[9] Grosheva, M.V., Efimov, G.B.: Computer analytical calculation systems. In: Applied programs packages. Analytical manipulations. pp. 5–30. Nauka, M. (1988)

[10] *Brumberg, V.A., Tarasevich, S.V., Vasiliev, N.N.: Specialized Celestial Mechanics systems for Symbolic Manipulations. Celestial Mech. 45(1-3), 145–162 (1988/1989)

[11] Klimov, D.M., Rudenko, V.M.: Computer Algebra Methods in Mechanical problems, p. 215. Nauka, M. (1989)

[12] *Computer Algebra in Physical Research. Int. Conf. Computer Algebra Phys. Res. Dubna, USSR (1990); Memorial Volume for N.N. Govorun, p. 453. World Scientific, Singapore (1991); Abstracts, p. 90. JINR, Dubna (1990)

[13] Abramov, S.A., Zima, E.B., Rostovzev, V.A.: Computer Algebra. Programirovanie (5), 4–25 (1992)

[14] *The History of Computer Algebra Applications. Session. In: 4-the International IMACS Conference on Applications of Computer Algebra, IMACS ACA 1998, Prague, August 9–11, 35 p (1998)

[15] Grosheva, M.V., Efimov, G.B., Samsonov, V.A.: Symbolic Manipulations in Control Theory. Izvestiya Akademii Nauk. The Control Theory & Systems (3), 80–91 (1998)

[16] Efimov, G.B., Yu, Z.E., Tshenkov, I.B.: Computer Algebra at Keldysh Institute of Applied Mathematics. Mathematical Simulation 13(6), 11–18 (2001)

[17] Grosheva, M.V., Efimov, G.B., Samsonov, V.A.: A History of Computer Symbolic Manipulation and its Applications in Mechanics. Keldysh Institute of Applied Mathematics of RAS, M., 87 p (2005)

[18] Dorodnitcin, A.A.: Solution of mathematical and logical tasks on computer. In: All-Union Confer., Moscow, March 12–17 (1956); Plenary reports. VINITI. M (1956)

[19] Okhotsimsky, D.E.: Research of motion in central field of forces under the influence of constant tangent acceleration. Cosmic Research. 2(6), 817–842 (1964); РЖМех, 11А29 (1965)

[20] Kantorovich, L.V.: On a mathematical symbolic system convenient for computer's calculations. Doklady AS USSR, 113(4), 738–739 (1957); Kantorovich, L.V., Petrova, L.T.: 3th All-Union Mathemat. Congress. Rep. (2), 151(1956); Smirnova, T.N.: Trans. Steklov Mathematical Institute. M. L. (1962); Pervozvanskaya, T.N., Ibid., Smirnova, T.N.: Polynomial PRORAB. Nauka, L. (1967)

[21] Stogny, A.A.: Solution on DCM of one task concerning differentiation of functions. In: Problems of Cybernetic, vol. (7), pp. 189–200. Nauka, M. (1962)

[22] Shurygin, V.A., Yanenko, N.N.: On a realization of algebraic differential algorithms on ECM. In: Problems of Cybernetic, vol. (6), pp. 33–43. Nauka, M. (1961)

[23] Brumberg, V.A.: Polynomial Series in Three Bodies Problem. Bull. ITA AS USSR, 9(4), 234–256 (1963); Brumberg, V.A.: Presentation of Planet's coordinates by trigonometric series. Reports of Institute Theoretical Astronomy (ITA AS USSR). L. (11), 3–88 (1966); Polozova, N.G., Shor, V.A.: Problems of motion of artificial Spacecrafts of Celestial bodies, M. (1963)

[24] Kabulov, V.K.: On deduce of differential equations of elasticity and construction mechanics. Doklady AS UzbekSSR, Tashkent, (9) (1963); Tolok V.A.: Problems of computer Math. & Techn. Tashkent (3) (1964); Kabulov, V.K.: Algorithmization in elasticity and deformational plasticity theory. "Fan" AS UsbekSSR, Tashkent (1966)

[25] Turchin, V.F.: Metaalgortmic Language. Cybernetika (4), 45–54 (1968); РЖМаТ, 2Б86 (1971); Turchin, V.F., Serdobolsky, V.I.: The REFAL and its application for transformations of algebraic expressions. Cybernetika (3), 58–62 (1969); Turchin, V.F.: Programming on REFAL. KIAM Preprint N. 41, 43, 44, 48, 49 (1971). РЖМаТ, 1В993 - 996 (1972)

[26] *Turchin, V.F.: Refal-5. In: Programming Guide and Reference Manual, New England Publishing Co., Holyoke (1989), http://refal.ru

[27] Zadyhaylo, I.B., Kamynin, S.S., Lyubimsky, E.Z.: Some problems of Computer design from blocks of high qualification. KIAM Preprint. M (1971)

[28] *Myamlin, A.N., Smirnov, V.K., Golovkov, S.L.: Specialized Symbol Processor. In: Woods, J.V. (ed.) Fifth Generation Architecture. New-Holland (1980)

[29] Smirnov, V.K.: Hardware REFAL realization at KIAM. KIAM Preprint N 99. M (2003)

[30] Zabrodin, A.V.: Super-Computers MVS-100, MVS-1000 and its applications to problems of physics and mechanics problems. Mathematical Simulation 12(5), 61-66 (2000)

[31] *Bruno, A.D., Edneral, V.F. Steinberg, S.: Foreword. Mathematics and Computers in Simulation. 45, 409-411 (1998); Bruno, A.D.: Normal Forms. Mathematics and Computers in Simulation. 45, 413-427 (1998); Soleev, A., Aranson A.B.: KIAM Preprint N 36 (1994)

[32] *Bruno, A.D., Varin, V.P.: The limit problems for equation of oscillations of a satellite. Celestial Mechan. and Dynam. Astron. 67, 1-40 (1997); KIAM Preprint N 124, 128 (1995) Sadov, S.Yu.: KIAM Preprint N 37 (1998)

[33] Kuleshov, A.S.: On the first integrals of Equation of motion of a Heavy Rotational Symmetric Body on a Perfectly Rough Plane. KIAM Preprint N 68 (2002)

[34] *Niukkanen, A.W., Shchenkov, I.B.: Operator factorization technique of formula derivation in the theory of simple and multiple hyper-geometric functions of one and several variables. KIAM Preprint N 81 (2003); A project of a globally universal interactive program of formula derivation based on operator factorization method. KIAM Preprint N 82 (2003)

[35] *Pogorelov, D.Yu.: Some developments in computational techniques in modeling advanced mechanical systems. In: Sympos. on Interaction betw. Dyn. & Control in Adv. Mechan. Systems, pp. 313–320. Kluger Acad. Publ, Dordrecht (1997); Pogorelov, D.Yu.: Differential-algebraic equations in multi-body system modeling. Numerical Algorithms 19, 183-194 (1988)

[36] Belousov, I.R.: Calculation of the Robot Manipulator Dynamic Equation. KIAM Preprint N 45 (2002); Balaban, I.Y., Borovin, G.K., Sazonov, V.V.: The language for programming right-hand sides of mechanical system motion equations. KIAM Preprint N 62 (1998)

Novosibirsk Young Programmers' School: A Way to Success and Future Development

Alexander Gurievich Marchuk, Tatyana Ivanovna Tikhonova,
and Lidiya Vasilyevna Gorodnyaya

A.P. Ershov Institute of Informatics Systems of SB RAS
Acad. Lavrentiev Ave, 6, Novosibirsk 630090
{Mag,tanja,gorod}@iis.nsk.su

Abstract. When in comes to school informatics, one certainly recalls the name of Andrei Ershov, whose energy and charm guided the first steps in the movement for computer literacy. In the mid-70's Ershov resolutely supported the founders of the Summer Schools of Young Programmers in Novosibirsk. Within the structure of the SB AS Computing Center's Experimental Informatics Department there was founded the first research subdivision with the purpose of formulating the conceptions and developing school informatics software – the Elementary Informatics Group. The guidance seminar "PC and the study process" was started in the Computing Center for the purpose of unification of education with the scientific potential. The ShYuP system, Shkolnitsa, the first textbook in informatics, the first computer study room in School N 166 in Akademgorodok made up of the Soviet-built Agat PCs were parts of a whole – the emerging and rapidly developing school informatics. The Summer School has a lot to be proud of. The Summer School raised and gave a start in life to a great number of programming specialists, who successfully work in the leading research institutes of the SB RAS, and computer and software companies such as Microsoft, Intel, and Excelsior. The early education in informatics, which showed its best in the intensive period of the Summer Schools, illustrates the multiple advantages of this approach. One cannot but mention the skill of naturally and softly use the human factor of working in a team while interacting with the outer world (what later became workshops), concentrating on teamwork under external orders. Another advantage was the chance to communicate with the whole world thanks to the many participants from other cities and countries (Summer Schools with international membership: Czech SSR, Bulgaria, Poland, DDR, France, Hungary). Summer Schools hosted people whose presence and company was a dream; John McCarthy's visit is an example. The pedagogical conception of the Summer School proved its opulence with an almost 35-year history. SB RAS Institute of Informatics Systems cherishes the traditions of Andrei Ershov's school and develops the successful experience of pre-professional programming education of talented youths.

Keywords: Computer science, systems programming, educational informatics, academician Ershov, summer school of programming.

J. Impagliazzo and E. Proydakov (Eds.): SoRuCom 2006, IFIP AICT 357, pp. 228–234, 2011.
© IFIP International Federation for Information Processing 2011

1 Formation

Since the period of academicians M.A. Lavrentjev and A.A. Lyapunov, much attention was paid to the Siberian schoolchildren who were considered as the main human resource of the Siberian science. The foundation of the physical-and-mathematical school, initiation of the All-Siberian mathematical contest, and the preparation of original teaching courses for secondary school all contributed much to discovering young talent.

Joint efforts of pedagogues, mathematicians, and programmers were aimed at creating an introductory course that should present a computer and basic notions of programming to pupils. During this work, two initiative groups have been formed. One of them was a group on computer science application of the Scientific Council on education problems of the Presidium of the Siberian Branch of the USSR AS. Another was a group of school informatics of the Computing Center of the Siberian Branch of USSR AS (CC); it was a subdivision of the experimental informatics department of CC especially organized for the elaboration of the concept and development of the software support for teaching informatics at school. A specialized seminar titled "Computer and teaching process" also started its work in support of integration between science and education.

2 School Informatics – The Beginning

Whenever we talk about school informatics, we always mention Andrei P. Ershov, a pioneer and leader in this field who headed the movement for "computer literacy". We can read in his diary [1] about the first contacts of pupils with the researchers of the programming department of the Institute of Mathematics of SB USSR AS in 1961, and in 1964, one of them became a teacher of programming in the 10th grade of one of the Novosibirsk schools.

The first lectures of the optional course of programming for pupils and their practical work on computers took place at the beginning of 1960-ties on the base of the school No. 10. The lessons were given by the researchers of the Institute of Mathematics (later they transferred to the Computing Center) with A.A. Baehrs among them.

Programming as a professional orientation course for 9-10th grade pupils was introduced into school No. 130 of the Novosibirsk Akademgorodok. We should note that this form of teaching became rather popular in Soviet schools and was in the school program until the 1990s. Several teaching courses were tested within these studies. Thus, at school No. 130, the courses were based on different programming languages such as Algol, Basic, and Fortran. It was also the place for experimental research in computer application to the school teaching process in general such as preparation of tests in English with the use of a computer-aided control system, which recognized the answers in a text form (this is now called the inter-subject relations).

3 Young Programmers' Schools

A.P. Ershov has strongly encouraged those who were the founders of Novosibirsk young programmers' schools. Among his first like-minded colleagues were N.A. Sadovskaya, the post-graduate of the Computing Center, and S.I. Literat, the head of the teaching department at school No. 130, the organizers of the first Young programmers' Summer School held in 1976.

Later, in 1977, A.P. Ershov invited G.A. Zvenigorodsky from Kharkov to participate in activity related to school informatics and, in particular, Young programmers' school (YPS). He was the author of the idea of the first programming environment for teaching process automation, (an applied program package "Shkolnitsa"), as well as the programming languages Robik and Rapira. A.P. Ershov considered the local YPS [2] as the most important form of working with schoolchildren. By 1981, there were about two hundred pupils in it.

Much attention was paid to the teaching plan for the junior (starting from the second grade) pupils. It should take into account not only the requirements to the programming languages but to the software system specially developed for teaching purposes. In addition to two lessons a week on programming theory, there was a possibility to work on computers of the Computing Center in the morning hours on Saturdays and Sundays. The teachers of the local YPS were researchers and engineers of Computing Center, and post-graduates and senior students of the Novosibirsk State University who were specialized in school informatics.

We should note that after a year of studies in YPS, many young programmers were involved in the real production projects. For example, they participated in the development of the system for analysis of primary structures of protein compounds, book exchange system, a tape perforation program for program-controlled embroidery machines, and so on. Besides, they took part in implementation of educational software systems, such as a compiler for Robik and Rapira, of the computer graphics system Shpaga and others.

In September 1979, a by-correspondence school for young programmers was started on the pages of the Kvant journal by N.A. Yunerman (Gein), a new member of the school informatics group just moved to Akademgorodok. Its best pupils were invited to Novosibirsk Summer YPS. Let us provide more details about the Summer YPS, a very important part of the programming teaching system.

4 Summer Young Programmers' Schools

The Summer Young programmers' school (SYPS) has many achievements for which to be proud. The system of informatics lessons for junior pupils clearly presents multiple advantages of this approach [3]. The by-correspondence school (namely, the contest held in the end of each school year) strongly influenced the selection of schoolchildren to be invited for two weeks to participate in SYPS.

After the first five years of holding SYPS, we had elaborated the system of studies. Pupils were separated into three groups: novices, those who knew the basic notions of programming, and young programmers who had some experience. It appeared actually that two weeks is quite enough for a novice to gain some skills in programming.

The curriculum of SYPS included two conferences. At the beginning, children presented their work implemented at home; the concluding conference was a report on programs written at SYPS. Some of presentations were recommended to be published in Kvant.

Among the favorite teachers of SYPS were well-known Soviet specialists and pedagogues like A.N. Terekhov, N.N. Brovin, L.E. Shternberg, O.F. Titov, Yu.I. Brook, and others who worked with enthusiasm on the first lessons of programming. Contacts with children of the same age and interests was very fruitful for everybody at SYPS, including its foreign participants from Czechoslovakia, Bulgaria, Poland, Germany, Holland, France, and Hungary. The visiting lecturers of SYPS were world-known persons, like John McCarthy, a famous American specialist in artificial intelligence. The SYPS had a great impact on the choice of future profession of a large number of its participants who work now in many Russian and foreign IT companies.

5 The Ershov Summer School of Young Programmers Today

Since 2001, SYPS is held during two weeks of July, a very convenient period for invitation of the NSU students and lecturers and the researchers of the institutes of SB RAS to work with schoolchildren. This event now takes place out of town in a tourist camping or a sanatorium in a picturesque place (once in two or three years they hold it in Altai [4]). The main goals of SYPS are the following: to select talented senior pupils interested in programming as their future job; to teach all the participants how to work in a team using modern IT technologies; to encourage junior pupils in their efforts to have more programming practice; to support teachers who are successful in this field.

The participants of SYPS are selected by their results shown at the previous Summer schools or at other events including the team contest in programming, the program "Young Siberian Informaticians", the contest in programming for the Novosibirsk region schools, the regional scientific-and-practical school conference in informatics and programming, and others. Some of the participants are personally invited by the teachers or by members of an organizing committee.

The members of the organizing committee of SYPS (many of them are researchers of the Ershov Institute of Informatics Systems of SB RAS) work much for the success of SYPS. Information about all abovementioned events is distributed through all possible channels including local newspapers and informative leaflets, the web-page of SYPS (at the IIS web-site), and booklets distributed at schools of Novosibirsk and other cities. Many pupils come from the cities of different Siberian regions, from Kazakhstan, Altai region and European part of Russia (St-Petersburg), even from foreign countries such as Norway.

As usual, SYPS is equipped with twenty-five to thirty computers provided by IIS SB RAS and sponsors. Unfortunately, they sometimes need substantial upgrade. Its participants in their photos and videos highlight school life. The most important breakthrough in the technical equipment of the school was the use of satellite telephony during the whole working period; it is very promising as a means of distant internet communication with those specialists who have no possibility to attend the school.

6 Workshop as a Basis of the Teaching Process

Since 1989, SYPS in Novosibirsk, in contrast to such schools in other cities, is held as a school with advanced courses in separate school subjects at pupil's choice, and its major goal is professional orientation of senior pupils. For this purpose, they introduce programming as an industrial activity with its problems, methodology, creative and technological aspects. The new notions and objects of studies are software product, the development process, correct problem statement and its formalization, job scheduling, debugging, preparation of documentation, and reporting on the project. For better comprehension of these notions, the teaching process is divided into ten or fifteen workshops of different profiles, according to different components of the software production cycle, where pupils can gain knowledge and skills during their teamwork on a project. The spectrum of topics studied in the workshops is very wide, so that every pupil could find something according to his level and interests.

The work in a workshop is supported with a course of lectures and specialized seminars on programming languages and systems, perspectives and problems in programming, informatics history and other disciplines, which provides pupils with a wide knowledge in computer science and interdisciplinary directions. In addition, there is an everyday "Task of the day" – a contest in solving an algorithmic task. A real pleasure for everybody is a cycle of lectures given by well-known scientists.

The major goal of each workshop is to go through all parts of a separate technological cycle in the framework of the stated problem and to present the results achieved in the process at the concluding school conference. The intensity of work needed to fulfill this task makes the stages of preliminary problem statement and job scheduling much more important. This was attractive for workshop supervisors in the aspect of testing new methods of project management under the permanent deficit of time and computers.

So, professional orientation of pupils of SYPS is directed to the following:

- Extend the knowledge of pupils in IT technologies and their applications;
- Present typical problems and methods to solve them,
- Present the sphere of possible future application of their capabilities,
- Provide pupils with special knowledge and skills of a team work.

7 Workshop Life

Preliminary division of pupils into workshops takes into account the results of interviews and their personal interests in a particular topic presented at the SYPS website. Children work in small groups supervised by experienced programmers on original projects. They become familiar with new computer tools and techniques and gain an invaluable experience of teamwork. The problem for a supervisor is not only to teach but also to create the atmosphere favorable for further growth of each participant of the project according to his capabilities, interests, and level. This level may vary but the necessary requirement is to know some programming language and to have certain programming skills. Workshops, in turn, are supervised by the head of the "teaching department".

To estimate the results of workshops' activity, there is a special jury of SYPS. The jury considers, first of all, the progress in pupils' knowledge and skills with the main attention paid to their capability to state a problem and to present and justify the results. Next, they estimate the software product qualities such as a convenient user interface, documentation, and presentation. Other parameters of estimation are the quality of report, a pupil's understanding of the task, and his or her own place in the workshop.

The members of jury give preliminary estimates a couple of days before the end of the school, which allows them to find weak places in the work and to make timely corrections in the teaching process. As usual, all workshops have time to complete demo-versions of their projects, and some of them prepare very good documentation.

The concluding conference of SYPS is keeping the best scientific traditions. The speakers are competent in their presentations, the audience is active and shows its high qualification, and true interest in the results here presented. It is a well-known fact that the decisive factor of success in any job is very often a good presentation of results, so participation in the concluding conference is obligatory for all workshops. All presentations are discussed and the best solutions are chosen. Many recommendations are given on improvement of other products. A very important person is the conference chair who is leading the discussion in a respectful manner.

8   Efficiency of the Summer Young Programmers School

The Summer School is efficient from many points of view. Thus, for the Siberian Branch of RAS, this is a very important mechanism for attracting young talents to computer science and the IT-sphere of national economy. For parents, this is a very good form of a summer recreation combined with obtaining knowledge-on-demand. Children coming from other places have a great opportunity to communicate with their new friends and just to have nice sightseeing.

For the Novosibirsk State University and, in particular, for its departments related to computer science and IT, the school made possible the following:

- Test the methods of teaching informatics to junior pupils,
- Attract to NSU more students interested in programming and IT who can take part in future contests and NSU projects,
- Raise the professional level of students,
- Attract students to teaching activity and to train them as future project managers and analysts (business-informatics).

For the administration of the Novosibirsk region, the experience of out-of-town work with children is very valuable because it can be extended and applied to rural schools, which will increase the quality of school education.

For IIS SB RAS, it is very important that its staff members and post-graduates take part in the experiment on teaching informatics in the form of a workshop, an idea that was suggested and implemented by researchers of this institute. They found that young people are interested in new forms of experimental work in the area of informatics systems. An experiment is in progress on organizing a programming school (Sundays, evenings, and distant learning) for advanced workshop teams during the school year; the theme "Investigation of informatics foundations and methods of teaching informatics and programming", which is in the research plan of IIS, is being developed. For the Novosibirsk IT-industry, the mechanism of SYPS provides an opportunity of early professional orientation of schoolchildren.

Almost thirty years of history of the summer school has proven that it is a consistent pedagogic idea. IIS SB RAS is carefully keeping up the traditions of Ershov's school and it is working for further progress in teaching informatics and programming.

References

[1] The Dictionary of the Head of Department, Academician A.P. Ershov's Archive, T. 35, p. 124

[2] Ershov, A.P., Zvenigorodskij, G.A., Literat, S.I., Pervin, J.A.: Work with schoolboys in the field of computer science. Experience of Siberian branch AN of the USSR, Mathematics at School (1), 47–50 (1981)

[3] The Report on the work of years school of young programmers, led from (August 14-28, 1987), on the basis of the tourist center Siberian
http://ershov.iis.nsk.su/archive/eaindex.asp?did=3936

[4] The Report on work of Years school of young programmers (2004),
http://school.iis.nsk.su/

[5] Marchuk, A.G., Tikhonova, T.I.: The workshop as the form of training to programming. Materials XV Intern. In: Conference on Information Technologies in Formation, Moscow, pp. 48–49 (2005)

"Lions – Marchuk": The Soviet-French Cooperation in Computing

Ksenia Tatarchenko

Princeton University, Princeton, New Jersey 08544-1017 USA
ktatarch@princeton.edu

Abstract. This paper looks to break with common assumptions of the underdevelopment and isolation of the Soviet computing by studying the history of the Soviet-French cooperation in computer science under the bilateral agreement of 1966. The achievements of this cooperation were largely due to the singular relations between French and Soviet mathematicians, J-L Lions and G. I. Marchuk. Although neither Marchuk nor Lions are computer scientists, properly speaking, they played crucial roles in promoting the use of computers as scientific instruments and in creating the administrative basis for the development of computer science. This work aims to present a trans-national history of computing and the networks, which existed between Soviet and Western computer scientists at the nexus of science and politics.

Keywords: Computers, Soviet Union, France, Cold War.

> *I have no doubt that communication is the principal way to understand the processes underlying the development of modern science.*
> G.I. Marchuk *Vstrechi i Razmyshleniia* (Moskva: Nauka, 1995), p. 9

1 Introduction

Two common assumptions dominate the perception of the history of Soviet computing: Soviet computers were few and not very good; levels of secrecy were extremely high, causing isolation from the international community.[1] This paper demonstrates that although these notions are not entirely incorrect, there is a case demonstrating that the development of Soviet computer science was on par with the international level, and that East-West cooperation in computing did exist. The case under investigation is the so-called "Lions-Marchuk" collaboration under the umbrella of the Soviet-French bilateral agreement of 1966. I attempt to answer the question: How did such cooperation become possible politically, institutionally and scientifically?

---

[1] Many studies are preoccupied by the question of technological independency, for example see: G. D. Crowe and S. E. Goodman, "S. A. Lebedev and the Birth of Soviet computing," *IEEE Annals of the History of Computing*, Vol. 16 (1994), no. 1: 4-24; and A. Fitzpatrick, T. Kazakova, and S. Berkovich, "MESM and the Beginning of the Computer Era in the Soviet Union," *IEEE Annals of the History of Computing*, Vol. 28 (2006), no. 3: 4-17.

J. Impagliazzo and E. Proydakov (Eds.): SoRuCom 2006, IFIP AICT 357, pp. 235–242, 2011.
© IFIP International Federation for Information Processing 2011

The three parts of the paper articulate the beginnings of Soviet-French cooperation in the following order. First, I introduce a vertical framework for the Lions-Marchuk relation, which is the Soviet-French Bilateral Agreement in Science and Technology. I argue that the agreement should be read in the context of the pivotal role that science and technology played in transatlantic hegemonic relations. In the second part, I examine the horizontal level of cooperation: How did the peculiarities of the development of computer technology and science in France and the USSR create the institutional bases and motivations for cooperation in computing? Finally, I study the realities of the so-called "Lions-Marchuk" cooperation.

2 *«Détente, entente et coopération»*

In March 1966, France left the leadership of the NATO, and, three months later, General de Gaulle undertook a triumphant official visit to the Soviet Union. This visit was, no doubt, a great event for the Soviet diplomacy. From the 20th to the 28th of June de Gaulle visited Moscow State University, Novosibirsk and its Akademgorodok, Leningrad, Kiev and Volgograd. The results of the visit are still a historical controversy. Some diplomatic historians assert that Soviets were left almost empty-handed: an invitation to visit Paris and a couple of scientific exchange agreements. The main issue of the day, the so-called "German Question," was left without any resolution. [1, 2] For some others, the French-Soviet Agreement signed on the 30th of June 1966 marked the French-Soviet *détente* and served as a model for the subsequent relations between the Soviet Union and Western countries. [13]

I believe that the scientific agreements between France and the USSR should be read in the context of John Krige's conclusion that basic science was the key node articulating American hegemony with the post-war reconstruction of science in Europe. [9] The French pursuit of nuclear capacity is the best-known element of the European resistance to American scientific and technological leadership. According to French technocrats the wartime defeat was to be overcome through technological development; the future of the French nation was at stake. [6] In this light, the agreements of the 30th June were more than a gesture and should be inscribed in a longer political movement of establishing French science and technology as independent from United States' influence.[2]

Whereas *informatique*, the French term for computer science, became international, early French computing history is sometimes told as a "history of technology transfer."[3] In fact, the French name for computer – *ordinateur*, – was invented in 1955 to better promote IBM machines on the French market. Furthermore, in 1964, the General was faced with a somewhat unanticipated set of problems. Bull, the only French company that specialized in computers for management, lost its majority holdings and became dominated by the American General Electrics

---

[2] For an earlier example of such a movement, see how the USSR adapted the French standard for its television: Walter Kaiser, "The PAL-SECAM Color Television Controversy," *History of Technology*, vol. 20 (1998), no.1: 1-16.

[3] Pierre Mounier-Kuhn, "History of computing in France," *IEEE Annals of the History of Computing*, vol. 11 (1989), no. 4: 237-240, p. 237. The whole volume is dedicated to the history of computing in France.

Company. France had an effective nuclear force de frappe, the ultimate demonstration of the triumph of De Gaulle's independent line; yet the failure to retain a native computing industry was significant. When in 1966 the USA put an embargo for a delivery of a big scientific computer (CDC 6600) ordered by France for its "bombe H" calculations, it triggered important governmental interventions in the affairs of French computing industry and science.

In the second half of 1960s, the new ambition of the French government was to challenge the American domination of the French computing market. The special inter-ministry commission created in 1965 rated informatics of "equal importance to atomic energy, civil aviation, and space."[4] On July 19th, 1966 General de Gaulle signed the *"Plan Calcul"* which prompted the creation of a new private computer company and a public research institute: CII (*Compagnie Internationale pour l'Informatique*) and INRIA (*Institut National de Recherche en Informatique et Automatique*). Symbolically, the newly created INRIA was attributed the location in Rocquencourt, a NATO military camp recently liberated from its occupants. Researches moved into the military caserns.

Within the Soviet Union the situation in the field of computing was rather complicated: a great deal of secrecy and contention split Soviet computing between the military and academia. [7, 12] The military benefited from a better material supply; meanwhile academia counter-balanced its scare resources with a relative openness. In October 1955, the existence of Soviet computers was revealed to the world at the now famous Darmstadt Conference, prior to being officially declassified for the Soviet Union's own citizens.[5] After this date there were rare visits and contacts between Soviet and Western computer scientists under the inter-academies exchange agreements and under the auspices of the International Federation for Information Processing (IFIP). [3] High level scholars connected with computing, such as academician Dorodnitsyn, were looking for every possibility of international contacts, and advanced the following reasons when arguing at the governmental level: to access easy sources for information collection; but also to prevent negative propaganda of the USSR's technological underdevelopment, provoked by insufficient demonstration of the Soviet achievements in computing.[6]

When in 1966 the French-Soviet Agreement provided a rare opportunity for cooperation, it was able to satisfy several Soviet interests: to grant the Soviets information on the French "state of the art" in computing, but also to provide a platform to access American know-how, since Americans were dominating the

---

[4] Henri Boucher, "Informatics in the Defense Industry," *IEEE Annals of the History of Computing*, Vol. 12 (1990), no. 4: 227-240, p. 238.

[5] Unfortunately, the history of the Darmstadt is relatively understudied, see: Slava Gerovitch, *From Newspeak to Cyberspeak: A History of Soviet Cybernetics*, (Cambridge, Mass.: MIT Press, 2002), pp. 155-58; Herman Goldstine, *The Compute from Pascal to von Neumann* (Princeton, NJ: Princeton University Press, 1972), pp. 349-62. For published sources see: "Darmstadt Conference Proceedings," *Nachrichtentechnische Fachberichte* (Braunschweig), vol. 4 (1956), no. 1; and Alston Housholder, "Digital Computers in Eastern Europe," *Computers and Automation*, no. 12:4 (1955), p. 8.

[6] A. A. Dorodnitsyn, "Justification of the USSR AS need to join IFIP", October 9, 1959 in Ershov Archive, see:
http://ershov.iis.nsk.su/archive/eaimage.asp?lang=2&did=28&fileid=78246

French market and French researchers had extensive connections with American scientists. Ironically, Akademgorodok, a center situated far away from Soviet Western frontiers, became the main beneficiary of this new possibility for cooperation. In 1958, academicians Lavrentiev, Sobolev and Khristianovitch acquired the support and funds from the Khrushchev government for the construction of a Siberian city of science. Designed as an alternative to the secret research centers working for the military (the infamous "boxes"), the new Akademgorodok was created to stimulate Siberian civil research, its connections with industry, and the exploitation of Siberian natural resources. Using his position at the head of a newly created Siberian branch of the Academy of Sciences and personal connections at the "top," Lavrentiev promised young talented researchers fast promotions and some cold but free air far from Moscow's sight. [8] These conditions rapidly made the scientific town an international center of great fame.

The context unfolded here can explain French and Soviet official willingness to cooperate in the field of computer science, endowing the agreement with some real content of "mutual interest." The international treaties that followed the visit of General de Gaulle to the USSR in 1966 provided the administrative and financial framework for cooperation. Cooperation between Soviet and French computer scientists functioned and developed according to these structures throughout the following decades up until 1993.

3   Akademgorodok CC – INRIA

INRIA, as a pilot institution for computer science, was responsible for Soviet-French cooperation in this field and represented France in all negotiations at the level of the working group in computing. Institutes specialized in computing under the auspices of the Academy of Sciences participated on the Soviet side. The Akademgorodok Computing Center (CC) became the main interlocutor of INRIA due to a constellation of circumstances: the official civilian character of the research done in Akademgorodok, the important role of computers as scientific tools for Akademgorodok sciences, and the human factor. Academicians Sobolev and Lavrentiev are known for their personal involvement with the early development of Soviet computing; unsurprisingly, they launched an initiative for the creation of a special CC in Akademgorodok.[7] Lavrentiev personally invited a young applied mathematician, Guriy Marchuk, to direct it. In 1964, the CC became independent from Sobolev's Institute of Mathematics and existed until the troubled times of nineties, when it was divided into several institutes.

The best account on the early history of Soviet computing under the umbrella of cybernetics is *From Newspeak to Cyberspeak: A History of Soviet Cybernetics* by Slava Gerovitch. The focus of the book is the complex relation between Soviet science and politics, which was responsible for the checkered nature and destiny of Soviet cybernetics. [5] Akademgorodok became home for Alexey Liapunov and

---

[7]   Notes on the conversation about the history of Soviet Computing with M. A. Lavrentiev by A. P. Ershov in Ershov Archve, see:
http://ershov.iis.nsk.su/archive/eaindex.asp?pplid=799&did=17909

Sergei Sobolev, two of three authors of the famous article-manifesto "Essential Characteristics of Cybernetics" [Osnovye cherty kibernetiki], which appeared in *Questions of Philosophy* [*Voprosy Filosofii*] in 1955 and marked a turning point in favor of cybernetics, previously publicly abused as pseudo-science. [14] Moreover, some new generation computer scientists, like Andrey Ershov, a talented pupil of Liapunov, gathered in Siberia at the very moment when Soviet cybernetics was gaining in power. Located at the periphery of the cybernetic movement, these scientists were involved with a different kind of computing, distinct from the interdisciplinary cybernetics. According to my interpretation the young Akademgorodok CC (newly created and staffed with a very young team) was implementing the principle "enough philosophy, lets get to work," and was intentionally seeking international contacts in order to build its reputation on a national scale. The quest for machines, budgets, buildings, and human resources was closely connected to the issues of scientific authority and fame.

The conditions in which the French INRIA had found itself were not entirely dissimilar. French scholars labeled the early period of INRIA's existence as "badly guided" – "*Institut malmené.*"[8] The challenges left to the young institute by the government, research for the French industry and education, were very difficult to realize in the conditions of insufficient and unqualified personnel, lack of machines, and a constant threat of decentralization. In addition, the heterogeneous administration constantly fought over the distribution of the tight budget. In order to assert its own identity, INRIA sought partners at home and abroad.

The first reports published by INRIA underlined its success in developing international connections. INRIA was responsible for education in computer science and distributed international fellowships, which helped a lot to establish initial contacts with the most prominent centers in the United States.[9] These first reports also revealed the role of INRIA in Soviet-French cooperation. The first president of the institute, Michel Laudet, traveled to Moscow to participate in French-Soviet working groups and set up the procedures for exchanges.[10] However, most interesting for us are the summary descriptions of the cooperation between the department of numerical computing directed by Lions and the Akademgorodok CC, directed by Marchuk. This line of collaboration became nicknamed "Lions-Marchuk cooperation," reflecting the importance of the personal relationship between the two scientists.

4 Lions – Marchuk Cooperation

In Marchuk's published memoirs, an almost legendary story tells us that, during his visit to Akademgorodok, General de Gaulle personally delivered J-L Lions' invitation to Marchuk to visit his laboratory in Paris.[11] The invitation in question was more likely made by Gaston Palewski, the Minister of Science and Technology from 1962

---

[8] Alain Beltran and Pascal Griset, *Histoire d'un pionnier de l'informatique* (Paris: EDP Sciences, coll. Sciences & Histoire, 2007), p. 17.

[9] *Rapport d'Activité de l'IRIA*, 1972, p. 73.

[10] *Rapport d'Activité de l'IRIA*, 1970, p. 59.

[11] G. I. Marchuk, O. N. Marchuk, *Neizvestnye stranitzy iz zhizni nekotorykh uchionykh* (Moskva: Nauka, 2002), pp. 137-38.

to 1965, during a different visit to Akademgorodok. However, the metamorphosis captured in the memoirs (published in 2002) transforms the Marchuk of 1966 into a hero with a historical mission and underlines the important political aspect of the beginning of the Lions-Marchuk cooperation.

Documents show that at first Marchuk visited Lions at the Institut de Blaise Pascal, part of the Sorbonne, in 1966. Soon Lions became the director of the department of Numerical Informatics, later renamed *Laboria*, the research unit at the heart of INRIA. Surrounded by the talented young people, so-called "*lionceaux*" ("lions' cubs" in French), Lions succeed in developing a powerful school of applied mathematics, and expanded considerably its field of operations. [4] I was told by Alain Bensoussan, one of the "*lionceaux*" who replaced Lions as head of INRIA, that long before any official cooperation Lions was particularly interested in Siberian splitting up methods and encouraged Bensoussan to employ them in his PhD research.[12] Marchuk also recalls being surprised by Lions' deep knowledge of his works; Lions read scientific Russian but couldn't speak it.[13] Apparently linguistic issues did not cause any trouble to establishment of mutual sympathy based on the commonality of research interests, but also on several less tangible factors, such as the tradition of collaboration between Russian and French mathematicians and common traits in personalities.

It is important to highlight that the collaboration worked out not only for Lions and Marchuk themselves: in fact, they also made their teams cooperate. The INRIA reports describe three principal themes of cooperation: 1) Techniques of decomposition, decentralization and parallelization (cooperation on the splitting up methods by N. Yanenko and R. Temam); 2) Identification of systems; 3) Optimization.[14] In brief, almost all of the most influential "*lionceaux*" participated in collaborations with Soviet scientists from Novosibirsk. Important personal links on the lower levels of the network were formed.[15] Such cooperation involved making their students work on adjacent topics and sharing their results. It was cumbersome to organize a long visit for a student or researcher, because of a difficult bureaucratic procedure, the request for a place from the so-called "long-stay" quotas. However, the government sponsored the short-term visits and annual meetings. From the end of the 1960s and during the 1970s, short visits were routine, and the so-called "Lions-Marchuk" colloquium took place every year, alternating between Akademgorodok and Paris. The main visible results of these meetings were publications, such as the following volumes: *Sur les méthodes numériques en sciences physiques et économiques*; *Étude numériques des Grandes systèmes*. [10, 11] The contents of these collections reflect 1970s tendencies within numerical analysis. The main topics of research were: principles of organization of

---

[12] Alain Benssoussan, personal communication to the author.

[13] Guriy Marchuk, personal communication to the author.

[14] *Rapport d'activité de l'INRIA*, 1969, p. 51.

[15] One of the favored examples of INRIA staff is a romantic story between a French researcher from INRIA, A. Morocco, and an interpreter from Akademgorodok. The love story grew into the marriage, witnessed by numerous documents in INRIA Archives. Morocco and his wife asked for research travel to Akademgorodok. Here is one quote from the correspondence between Lions and Marchuk: "...you already know him. He would go with his wife, who is from Novosibirsk (the best example of collaboration!!!)" January 10, 1978, INRIA Archives, box 83.06.035.

software systems for the resolution of partial differential equations by finite elements; modeling and optimization of complex systems; and technical aspects of implementation of methods of approximation using supercomputers.

One important factor that both Lions and Marchuk had to deal with was the well-known problem of the lack of machines. They were interested in developing numerical techniques allowing optimization of machine power. INRIA and Akademgorodok CC researchers had to proceed with a different philosophy compared to the American approach when resolving scientific problems. There is a direct statement disclosing these differences of practices in Lions' letter to Harold Agnew, director of the Los Alamos Laboratory from 1970 to 1979. Lions explains: "We have to work in advantage on the models and mathematical aspects before going to the computer. From this point of view (and only from this one!) we work more like the Russians."[16] The oral history interviews I conducted with Lions' collaborators transmitted a very similar message, but without the parenthetical note.[17] Two schools of numerical analysis, the French and the Siberian, had at least two characteristics in common: they had "to think more" (than Americans) and they relied heavily on personal networks, structured around the father figures of the school-founders.

Although neither Marchuk nor Lions are computer scientists, properly speaking, they played crucial roles in promoting the use of computers as scientific instruments and in creating the administrative basis for the development of computer science. Both Lions and Marchuk were sought-after and successful advisers, keeping links with their former students and actively helping them to acquire positions, so that the "Lions-Marchuk cooperation" acquired the character of "networks in cooperation." Both Lions and Marchuk not only frequently crossed national frontiers, but also circulated with ease in the different spheres: science, politics and industry. From the middle of the 1970's and during the following years the careers of these two men underwent a dramatic rise. They became powerful administrators of the "big science" at the highest levels.[18] A consequence was a change in the nature of their involvement in cooperation matters.

5 Conclusion

We saw that for the cooperation between the French and Soviet computer scientists to exist, an accumulation of political, institutional, and personal willpower and interest was indispensable. The French government wanted to cooperate in order to build up a computer industry and research independently of American efforts. The Soviets were trying to access the Western computing world. Both INRIA and Akademgorodok CC were young institutions and used international contacts to create their own identity and to establish their authority within the national context. Finally, Lions and Marchuk were the human engines of the cooperation. They used the political agenda

---

[16] J-L Lions to Harold Agnew, letter quoted in Dahan-Dalmedico, *Jacques-Louis Lions*, pp. 183-84.

[17] George Nissen, personal communication to the author.

[18] Marchuk is best known as the last president of the Soviet Academy of Sciences and Lions left INRIA to preside over the French national center for space studies (CNES).

to realize their scientific and career ambitions, generating transnational links at the different levels of their personal networks.

The Lions-Marchuk cooperation demonstrates that Soviet –West cooperation in computing did exist under certain political circumstances, and that the well-known Soviet shortage of machine power contributed to the development of numerical techniques of great interest to Western researchers. Finally, this paper reveals the potential promise of looking beyond the Soviet-American antagonism.

Acknowledgments. This paper is based on my master thesis defended at the Sorbonne in June 2005, *INRIA – Akademgorodok, a history of French-Soviet Cooperation in Computer Science, 1966-1993*. I would like to thank the people who granted me interviews and supplied other information, particularly G.I. Marchuk, A.S. Alexeev, V.P. Orechenko, A.B. Ugolinokov, A. Bensoussan, G. Nissen, and P. Nepomyastchy.

References

1. Bariéty, J.: Le voyage de de Gaulle à Moscou en 1966 et la question allemande (D'après les documents des Archives de ministère français des Affaires Etrangères). In: Rossiia i Frantziia, XVIII-XX veka. vypusk, vol. 4, pp. 282–289. Nauka, Moskva (2001)
2. Belousova, Z.S.: K vizitu generala de Gollia v Sovetskiy Soiuz v 1966 godu (Na osnove arkhivov Ministerstva Mezhdunarodnykh Del Rossii). In: Rossiia i Frantziia, XVIII-XX veka. vypusk, vol. 4, p. 290–295. Nauka, Moskva (2001)
3. Carr III, J.W., Perlis, A.J., Robertson, J.E., Scott, N.R.: A Visit to the Computation Centers in the Soviet Union. Communications of ACM 2(6), 8–20 (1959)
4. Dahan-Dalmedico, A.: Jaques-Louis Lions, un mathématicien d'exception entre recherche industrie et politique. La Découverte, Paris (2005)
5. Gerovitch, S.: From Newspeak to Cyberspeak. MIT Press, Cambridge (2002)
6. Hecht, G.: The Radiance of Franc: Nuclear Power and National Identity after World War II. MIT Press, Cambridge (1998)
7. Ichikawa, H.: Strela-1, the First Soviet Computer: Political Success and Technological Failure. IEEE Annals of the History of Computing 28(3), 18–31 (2006)
8. Josephson, P.: New Atlantis Revisited: Akademgorodok, the Siberian City of Science. Princeton University Press, Princeton (1997)
9. Krige, J.: American Hegemony and the Postwar Reconstruction of Science in Europe. MIT Press, Cambridge (2006)
10. Lions, J.L., Marchouk, G.I. (eds.): Méthodes Mathématique de l'Informatique – 4, Sur les méthodes numérique en sciences physiques et économiques, Paris (1977)
11. Lions, J.L., Marchouk, G.I. (eds.): Méthodes Mathématique de l'Informatique – 7, Etude numérique des grands systèmes, Paris (1978)
12. Malinovskii, B.N.: Istoriia vycheslitel'noi tekhniki v litzakh. Naukova Dumka, Kiev (1995)
13. Rey, M.P.: La Tentation du rapprochement: France et URSS à l'heure de la détente, 1964-1974. Publications de la Sorbonne, Paris (1991)
14. Sobolev, S.L., Kitov, A.I., Liapunov, A.A.: Osnovnue cherty kibernetiki. Voprosy filosofii (4), 136–148 (1955)

Information and Communication Technology Education Based on the Russian State Educational Standard of "Applied Mathematics and Informatics"

Iurii A. Bogoiavlenskii

Petrozavodsk State University, Petrozavodsk, Republic of Karelia, 185910, Russia
ybgv@cs.karelia.ru

We strive in life so in the end
Paradise attain, to heavens ascend.
It is better our ways amend
This moment now, joyously spend.

– Rubaiyat of Omar Khayyam
English translation by
Shahriar Shahriari

Abstract. The expansion of the computing field seriously challenges academic planning, as it requires acceleration of new curricula development and implementation. Academia possesses much inertia and curricula changes are slower than the computing expansion. American computer societies under the aegis of ACM and IEEE have developed separate curricular guidelines for disciplines in computer engineering, computer science, information systems, information technology, and software engineering. Our analysis of the guidelines shows that they contain a large amount of mathematical courses. We offer a "reverse" approach to curricular guidelines development where we introduce engineering components in the existing curriculum with intensified mathematical training. In the Russian Federation, we may use the "Applied Mathematics and Informatics" standard for these purposes. Our comparative analysis of body of knowledge cores for four disciplines (except computer engineering) shows that these core areas are entirely accommodated in the core of the Russian standard. We believe that mathematical culture provides long-term efficient professional activity of graduates.

Keywords: Computing curricula, Intensified mathematical training, body of knowledge cores, Russian standard.

1 Introduction

In this article, we will use the term "information and communication technologies" (ICT) as an analogue to the term "computing" as accepted in the United States. The ICT field is rapidly expanding; governments and private companies invest in its development vast sums of money. New paradigms, concepts, standards, tools, and application systems appear and deployed very quickly, supplementing and/or replacing each other.

J. Impagliazzo and E. Proydakov (Eds.): SoRuCom 2006, IFIP AICT 357, pp. 243–250, 2011.

Considering this process, the authors of [1] note that the 1990s witnessed ICT specialist diversification, which resulted in the formation of five family disciplines: computer engineering (CE) [2], computer science (CS) [3], information systems (IS) [4], information technology (IT) [5], and software engineering (SE) [6].   This development considerably increases both importance and complexity of adequate curricula models development and deployment problems to accomplish the task of grounding specialists capable of long-term efficient professional activity in ICT fields. This problem is constantly the focus of attention for ICT communities in the United States [1-7], in Russia [8-14], and worldwide [15] as well.

In [1] the committee proposed to have separate curricular guidelines of undergraduate programs for each of the five disciplines. The purpose of this article is to show that it is possible (and expedient) to organize efficient education of specialist for the four disciplines CS, IS, IT, and SE in the framework of the Russian state educational standards family "Applied Mathematics and Informatics" [16]. One may deploy such an education mainly based on baccalaureate direction 010500 ("old" cipher 510200), followed (if necessary) by a specialization.

2   Motivation to Use the Standards Family "Applied Mathematics and Informatics"

New disciplines curricular guidelines development is undoubtedly an important task. At the same time, institutions of higher education possess much inertia and the speed of introduction and stabilization of new guidelines they can provide is considerably lower than the rate of ICT changes viewed nowadays. Under existing conditions, one of the sensible reactions aimed at accomplishing the task of infusing new curricula guidelines is the adjustment of the existing standards that are widely spread within the universities.

Flexibility is one of the important ICT curricular guidelines requirements.   It allows successful use of them under diversified conditions for a rather long period. It provides flexibility by having in a standard a permanent part serving as a basis and a variable part for adjusting to the changes. To form the permanent part of the standard in Russia, it is natural to rely on Russian tradition to provide education fundamentals. The foundations of the qualification allow the production of a graduate of sound quality with a high probability of professional productivity. The authors of [3, section 9.1.6.] share the approach.

The role of mathematics in ICT formation and development is a fundamental one, as the ICT professional deals with formal and abstract concepts, and objects. In [3, part 9.1.1.] it stresses, "Mathematics techniques and formal mathematical reasoning are integral to most areas of computer science. <...> Given the pervasive role of mathematics within computer science, the CS curriculum must include mathematical concepts early and often". Language in [4-6] expresses the same point of view.

Moreover, CS curricula models for the USA research universities include "from a one-semester course to a sequence with three or more courses" [3, part 9.4.1] of calculus. (Note that the 010500 standard provides from three up to five terms of calculus.)   The list of in-depth courses [3, part 9.3.] addresses subjects as combinatorial analysis, probability and statistics, and many other topics, which are obligatory in the 010500 standard.

All of this allows for a sound formulation of a highly important statement that applied mathematical methods form the basis of the CS discipline. Therefore, the most important requirement to an ICT discipline curricular guideline is to form integral mathematical culture (MC) of a graduate. Possessing it, a graduate will quickly master any contemporary and coming ICT concepts, methods and technologies that are of critical importance under rapid ICT changes observed. Note also that possessing MC will enable students to master engineering ICT components quicker.

The forecast of ICT diversification process development is not clear. One of the rather likely scenarios is that due to monopolization, concentration, the applied side of ICT will reach a point of simplification, and the diversification process will begin to go in the reverse direction.

We propose, under the ambiguity conditions caused by diversification, to use the "reverse" approach to curricular guidelines formation. When we include the corresponding engineering constituents in a guideline, it will require a rather complete mathematical education. Firstly, this approach fits perfectly within the fundaments of education and accommodates the flexibility of curricular guidelines. Secondly, it provides faster reaction on diversification by adjusting existing widespread and stable university standards.

3   Characterization of Standards Family 01050{0|1} - "Applied Mathematics and Informatics"

In Russia there exists since 1993 a state educational standards family [16] "Applied Mathematics and Informatics" with education in specialty 010501 (Specialists - five years) and in the directions of 010500 (Bachelors of Science - four years, Masters of Science - six years). The bachelor standard has the following basic characteristics. The total student work content is 7,314 hours, which accounts for 132 weeks of study (eight terms). Standard courses fragment by blocks, shown as follows.

o   Humanitarian and socio-economical sciences - 24%
o   Natural sciences - 9%
o   Mathematics general - 28%
o   Applied mathematics - 16%
o   ICT general - 12%
o   ICT according to the faculty decision and elective courses - 11%

Therefore, it is possible to provide about 1,700 hours of study of ICT within the framework of the standard; that comprises 23% of the total student work time, which programs can increase if necessary up to 30% (2,190 hours) due to the permissible standard variations.

The specialist education standard 010501 provides a total of 8,032 hours of student work content over nine terms; the tenth term is predefined for practical training and degree work development. Standard content and structure are in close agreement with the 010500, bachelor standard; the work content is a 718-hour increase compared to the latter. Therefore, the fifth year of studies allows ICT study time to increase up to 2,418 hours (or 30% of the study hours).

The master training standard 010500 provides a total of 4,100 hours of undergraduate work content during 88 weeks of study for two years after completing the baccalaureate degree. Standard flexibility provides twelve different training sub-areas (problem fields). The ICT sub-areas are 010509 for software design, 010510 for networks software, and 010511 for system programming. The standard provides the use of 73% of the study hours for undergraduate mastering in the chosen ICT sub-area. The standards family has proven itself beneficial in Russia [9]; it provides integral MC formation, sufficient time for basic ICT courses, and an elective block to reflect current ICT changes.

We now examine a question whether the 010500 bachelor standard has enough study hours to accommodate the core subjects of CS, IS, IT and SE disciplines' bodies of knowledge (BK) specified in [3-7].

4 Comparative Temporal Characteristics of Bodies of Knowledge Cores

We have conducted a comparative research on study hours, provided for BK cores for the IS [4, 7], the IT [5], and the SE [6] disciplines concerning CS BK Core [3]. The comparison with a description appears in Tables 1 and 2 in an extended version of this article, located at URL http://www.cs.karelia.ru/news/2006/files/sorucom-ybgv-en.pdf. A summary of these tables appears in Table 1.

Table 1. Summary of Discipline Comparison

	Hours			
	CS	IS	IT	SE
Total lectures in common part of core	286	278	143	250
Lectures in common part of core (Without mathematical Subject Area Discrete Structures)	243	235	100	207
Lectures in special parts	N/A	245	181	244
Lectures in core	243	480	281	451
Study time in special parts	None	980	724	976
Study time in core	972	1920	1124	1804

Later on, we will use the "B standard" term for the phrase "010500 Bachelor standard". With this standard, designate as B-ICT, the study hours related to ICT study total 1,920 hours.

The temporal analysis shows that the CS core is almost entirely included in IS, SE and IT cores. Thus, we have come to the fundamental conclusion that CS core knowledge is basic for the general ICT sphere; that is, they are fundamental for the IS, IT, and SE disciplines, which implies that the latter to be applied disciplines relative to CS. This conclusion agrees well with sections of [4-6] devoted to IS, IT and SE cores connections with CS core. We also see, that the B standard, considering the discrete structures area as a natural introduction in its general mathematics courses block, allows accommodating the total study hours of the BK cores extend freely to each of four disciplines in B-ICT.

Finally, the temporal comparative analysis allows us to draw the following conclusions on total study hour correlation provided in B standard and in the guidelines [3-7] for CS, IS, IT and SE disciplines.

A. In [3] it states that the total CS study hours in American and Canadian universities vary greatly and it does not mention its precise extent. When computing it, assuming that besides the core the graduate is to study 12 CS courses at 160 hours each; all in all (except for the discrete structures subject) it gives 237 lecture hours in the special part in addition to 243 core hours. That is, it contains 480 lecture hours or 1,920 study hours. It means that total CS study hours (not only BK) is entirely accommodated in B-ICT. Thus, we can think of the B standard to be a guideline [3] equivalent in both content and CS knowledge extent.

B. The total IT study hours (1,800 lecture hours) are not only entirely accommodated in B-ICT, but they allow (if using B standard for IT training) the recommended mathematical training [5, section 8.1.1] and advanced courses [5, section 8.2] for graduates.

C. The IS core extent is also accommodated in B-ICT. It appears in [4]. That is, the core extends for two years of study and the recommendations on total study hours are given in a general form as, "Prerequisite or interleaved topics directly applicable to the IS curriculum therefore include: <...> discrete mathematics, introduction to calculus, introductory statistics, <...> principles of economics and functional areas of the organization such as accounting, finance, human resources, marketing, logistics <...>". When using the B standard to train IS specialists, the study of mathematical courses occurs naturally, and economic and organization courses that are not in IS core, can be offered at the cost of humanitarian time block and partly in the applied mathematics block of the B standard.

D. The SE Core is accommodated entirely in B-ICT. Additional courses, recommended in [6] for the complete curriculum introduction (calculus, physics, humanitarian, and social courses) belongs also to B standard. Thus, the considerable part of total SE study hours can also be accommodated in the B standard.

The question about the possibility of accommodating the total study hours of IS and SE disciplines in B standard lies outside the scope of this article and requires additional research. At the same time it is possible to say with certainty that the specialist education standard 010501 (five years of study) allows all four disciplines graduate training in total compliance with recommendations [3-7].

Standard B adjustment for the four graduate ICT disciplines becomes a fixation of three professional blocks: the mathematical core, the ICT core, and the special blocks corresponding to each of the disciplines.

5   The Bachelor Direction 010500 Use Experience in Mathematical Faculty of Petrozavodsk State University

In Petrozavodsk State University, the 010500 (510200) Bachelor direction was open in 1993 [17-19] and Master of Science direction in 1997. We developed the curriculum by taking into account the Computing Curricula 1991 recommendations. Over thirteen years, the faculty has graduated 269 Bachelors of Science students, 191 Specialists, and 73 Masters of Science students. Only a few students left the university to continue their Bachelor of Science degree at other universities. The majority of the students continued their education for one or two years to obtain Specialist or Master of Science Diploma. Graduates frequently continued their education in postgraduate school.

Our experience verifies that the B standard provides exceptional flexibility enabling one to reflect current ICT changes in the curriculum. For the last thirteen years, we have successively introduced in the curriculum studies such as introduction to processors, computer networks, operating systems, software engineering, shell language, object-oriented programming in Java and .NET environments, web technologies, and SE team project courses.

Due to the elective and facultative courses, specialized training for the system network technologies program was open in 2001 where the following courses are offered: concurrent systems, OS Unix programming, network programming, and distributed systems.

6   Conclusion

We have examined educational standards development and deployment problem to accomplish the task of specialist training capable of long-term efficient professional activity in the information and communication technologies sphere under its diversification and rapid considerable changes conditions. The article offers to serve the growing needs for the specialists by the adjustment of existing, widely spread standards that significantly decreases the university response time to the ICT sphere of content and demand changes.

Reasoning from the thesis that applied mathematical methods form the basis of the computer science discipline, we emphasize the importance of mathematical culture for all ICT disciplines and offer a "reverse" approach to curricular guidelines development. The gist of the approach is to introduce corresponding engineering components into a curriculum that provides intensified mathematical training. As a basic curricular guideline in Russia, it is natural to use the 01050{0|1} standard family; i.e. the "Applied Mathematics and Informatics", which is widespread in Russian universities and has been successively approved for a long time.

Accomplished temporal characteristics comparative analysis of "Computer Science", "Information Systems", "Information Technologies" and "Software Engineering" disciplines Body of Knowledge Cores shows that these Cores are entirely accommodated in the study hours of Bachelor direction 010500 provided for ICT study.

The 01050 {0|1} standard family adjustment approach to contemporary needs and ICT sphere state has the following advantages:

o   Directed the integral Mathematical Culture formation;
o   Approved the two stage schema for the Bachelor of Science and Master of Science programs;
o   Flexible structure providing easy reaction to ICT changes;
o   Adjustment process procedure simplicity compared to development, approbation, and new standards introduction processes.

The 01050{0|1} standard family uses experience in Mathematical Faculty of Petrozavodsk State University during thirteen years justifies the thesis offered in the article.

References

1. IEEE/AIS/ACM Joint Task Force on Computing Curricula. Computing Curricula 2005. The Overview Report covering undergraduate degree programs in Computer Engineering, Computer Science, Information Systems, Information Technology, Software Engineering (2005), http://www.computer.org/curriculum, http://www.acm.org/education/curricula.html
2. IEEE/ACM Joint Task Force on Computing Curricula. Computer Engineering 2004. Curriculum Guidelines for Undergraduate Degree Programs in Computer Engineering. IEEE Computer Society Press and ACM Press (2004), http://www.computer.org/curriculum, http://www.acm.org/education/curricula.html
3. ACM/IEEE-Curriculum 2001 Task Force. Computing Curricula 2001, Computer Science. IEEE Computer Society Press and ACM Press (2001) http://www.computer.org/curriculum, http://www.acm.org/education/curricula.html
4. ACM/AIS/AITP Joint Task Force on Information Systems Curricula. IS2002 Model Curriculum and Guidelines for Undergraduate Degree Programs in Information Systems, Association for Computing Machinery, Association for Information Systems, and Association for Information Technology Professionals (2002), http://www.acm.org/education/curricula.html, http://www.computer.org/curriculum
5. The ACM SIGITE Task Force on IT Curriculum. Information Technology, Computing Curricula Information Technology Volume. Curriculum Guidelines for Undergraduate Degree Programs in Information Technology, http://www.acm.org/education/curricula.html
6. IEEE/ACM Joint Task Force on Computing Curricula. Software Engineering 2004. Curriculum Guidelines for Undergraduate Degree Programs in Software Engineering. IEEE Computer Society Press and ACM Press (2004), http://www.computer.org/curriculum, http://www.acm.org/education/curricula.html
7. Detailed Body of Information Systems Knowledge, http://192.245.222.212:8009/IS2002reportsPDF/rptBodyOfKnowledge.pdf
8. Terehov, A.N.: How to train system programmers. Computer Tools in Education (3-4), 2–80 (2001) (in Russian)

9. Ivanovski, S.A., Liss, A.R., Romantsev, V.V., Ekalo, A.V.: Programmer's Training within the framework of State Education Standard Major and Specialization. Presentation at the 2nd conference on Information Technologies Instruction in Russia (2004) (in Russian), http://www.it-education.ru/2004/reports/romantsev.htm

10. Suhomlin, V.A.: IT Education: Concept, Educational Standards, Standardization Process. Hot Line - Telecom, Moscow (2005) (in Russian)

11. Nikitin, V.V.: ICT Education System Standards and Structure Development. Presentation at the 3rd Conference on Information Technologies Instruction in Russia (2005) (in Russian), http://www.it-education.ru/2005/reports/1_Nikitin.htm

12. Proceedings of First International Scientific and Practical Conference on Contemporary Information Technologies and IT Education. Moscow State University Publishing House (2005) (in Russian)

13. State Higher Education Standards: 010503 "Information Systems Software and Administration", 010300 "Mathematics, Computer Science", 010400 "Information Technologies", 080700 "Business Informatics". Ministry of Education of Russia. Moscow (2000 - 2005) (in Russian)

14. Information Technologies Instruction in Russia Conference resolution (2005) (in Russian), http://www.it-education.ru/2005

15. Goldweber, M., Impagliazzo, J., Clear, A.G., Davies, G., Bogoiavlenskii, I.A., Flack, H., Mayers, J.P., Rasala, R.: Historical perspectives on the computing curriculum (Report of WG no. 7). Working Group Reports and Supplemental Proceedings of ITiCSE 1997, New York, USA, pp. 94–111. ACM Press, Uppsala (1997)

16. State Higher Education Standard. 010500, Applied Mathematics and Computer Science. Degree - Bachelor of Applied Mathematics and Computer Science, 010501 - Applied Mathematics and Computer Science. Qualification - Mathematician, system programmer, 010500, Applied Mathematics and Computer Science. Degree - Master of Applied Mathematics and Computer Science, Ministry of Education of Russia, Moscow (2000) (in Russian)

17. Voronin, A.V., Bogoiavlenskii, Y.A., Kuznetsov, V.A., Poljakov, V.V., Sigovtsev, G.S.: E02 "Applied Mathematics and Computer Science" Baccalaureate Education Program. In: All Russian Guidance Conference "Teachig strategy bases of Multilevel Education System Development and Functioning" abstracts (in Russian)

18. Pechnikov, A.A., Bogoiavlenskii, Y.A., Voronin, A.V., Kuznetsov, V.A., Poljakov, V.V., Sigovtsev, G.S.: E02 "Applied Mathematics and Computer Science" PetrSU Baccalaureate Program. Applied Mathematics and Computer Science, vol. 3, pp. 75–80. Petrozavodsk State University Publishing House (1994) (in Russian)

19. Bogoiavlenskii, I., Pechnikov, A., Sigovtsev, G., Voronin, A.: Using of Computing Curricula 1991 for Transition from "Mathematics" to "Applied Mathematics and Computer Science" Baccalaureate Program. In: Abstracts of Conference ITiCSE 1997, p. 8. University of Uppsala, Uppsala (1997)

Cooperation among Institutions of the Soviet Union and Cuba: Accomplishments between 1972 and 1990

Tomás López Jiménez

University of Informatics Sciences
Carretera a San Antonio de los Baños Km. 2½, Torrens, Boyeros
Ciudad de la Habana, 19370, Cuba

Abstract. This work provides a landscape on the manner in which Cuba and the Soviet Union had cooperated in the computing field between 1972 and 1990. It highlights some of the important milestones between the two nations including Cuban membership on the Council for Mutual Economic Assistance, the COMECOM, and its activities in the Coordinating Committee for Multilateral Export Controls. It also addresses Cuba's relationship with the Former Soviet Union after 1990.

Keywords: Computing in Cuba, Soviet Union, COMECOM, CMEA.

1 Introduction

The SoRuCom 2006 Conference has offered a unique opportunity to exchange information on contributions to electronic computing (EC) from the Former Soviet Union (FSU). These facts, both worthy and significant for students, professors, and professionals of its disciplines worldwide, need further documentation by the history of computing.

This conference occurs sixty years after the public demonstration of the ENIAC computer, immediately after World War II and three years before the establishment of the Council for Mutual Economic Assistance (CMEA) in 1949, under the leadership of the FSU. The intermingling of causes and effects among these events, their true motivations, and their usefulness for humanity are still partially and barely researched or informed. In fact, the effects induced in the CMEA countries by the multilateral collaboration constitute part of the history of the scientific development of the EC promoted by the FSU, which dedicated special efforts to this collaboration; thus, empowering it with the creation in 1969 of the Intergovernmental Commission for Collaboration in the field of EC (ICCEC). Therefore, part of the accomplishments of the joint works is also part of Soviet and Russian computing history.

Cuba became a member of CMEA in 1972, three years after the institution of the ICCEC, over twelve years from the beginning of bilateral relations with the FSU. This paper discloses some elements of this bilateral history on EC. First, we explore its principal landmarks in Cuba since 1972; then we also reflect on some of the main achievements from the rich and friendly Cuban-Soviet collaboration in this field.

J. Impagliazzo and E. Proydakov (Eds.): SoRuCom 2006, IFIP AICT 357, pp. 251–257, 2011.
© IFIP International Federation for Information Processing 2011

2 Development of the EC in Cuba Before Its Admission to COMECOM

2.1 Computing in Cuba Before 1 January 1959

When the Cuban Revolution triumphed on 1 January in 1959, the knowledge of EC did not exist in a practical sense in the country. This was the status even though there was a vast tradition using tabulating and accounting machines for data processing, including the use of some electronic ones since 1957 that included two IBM 407 machines and a UNIVAC 120. Actually, some private schools, mostly Cuban, prepared specialists for their use.

By the end of 1958, the first electronic computer appeared in Cuba; it was a first generation IBM RAMAC 305 with the first magnetic discs in the world, when they still constituted a world novelty [1]. One should note that in 1927 IBM opened a business office in Cuba, the 27th outside the borders of the USA; in fact, it placed in the 16th position among its main landmarks of foreign business concerning the installation of tabulators in Havana in 1927. Cuba was the thirteenth country to which they exported their equipment [2]. The Cuban government confiscated IBM Cuba in 1961. These facts are evidence that in 1959 there were in Cuba some advanced components of EC. In addition, the minimum necessary conditions for starting its development were already in operation and the country exerted a certain attraction for IBM [3]. When the United States government broke its diplomatic relations with the Cuban government in 1961, it imposed a ferocious blockade on the country that started in February of 1962, thereby making access to their technology on the island very hard or almost impossible.

2.2 First Steps: The 1959-1967 Period

In 1959, a deep educational reform started in the country. In January of 1960, Commander-in-Chief Fidel Castro stated, *"The future of our country has to be necessarily a future of scientific men, a future of thinking men."* Later on in March of 1962, the Minister of Industries, Commander Ernesto Che Guevara pointed toward electronics as one of the four important lines for the development of the country. That same year the government organized the Automation and Electronics Division with the mission of developing research and working experiences in electronics, cybernetics, automation, and computing [4].

In 1964, the National Calculus Center (CNC) at the University of Havana (UH) acquired the British second-generation computer – the Elliott 803B; the university completed its installation in 1965. The School of Mathematics and other areas of the university, in collaboration with the Department of Applied Mathematics of the Central Planning Board (JUCEPLAN), as well as other institutions, developed applications for the optimization of transportation, diets optimization and management of chicken plants, numerical weather forecasts, and other areas that benefited the country.

The sugar industry developed important applications for the period as part of its perspective plan to produce ten million tons of sugar in 1970. It all started with the RAMAC and until 1966, it progressively used the Elliot 803B in applications with regression and linear programming models in order to optimize the harvest period and the composition of sugar cane varieties per plant, the operational control, as well as the optimization of the transportation of the sugar cane among other areas.

2.3 The Essential Conditions Were Created: The 1968-1971 Period

Between 1968 and 1976, the following parallel and complementary paths directed the Cuban development of the EC:

1. Under the responsibility of the JUCEPLAN, the board fostered planning and statistical centralized control over the country plans and programs and the 1970 population and housing census. The National Calculus Plan (PNC) was created for this program, based on the importation of technology and technical assistance. In 1969, the Electronic Calculus Division was also organized.
2. The second path foresaw empowering the scientific and technological national development. Personally conceived and encouraged by Commander-in-Chief Fidel Castro by the end of 1968, it was assigned to the UH, with an immediate objective of developing the first Cuban computer, a real challenge for a very young scientific policy [5].
3. The third path was under the responsibility of the national defense and security organisms; it also included a tight collaboration with the other two.

The PNC reached agreements in 1968 with the French government in 1970 for the supply of two second-generation SEA 4000 computers. From 1972 on, there were other supplies of two mainframes IRIS 50 and some minicomputers IRIS 10, both were third-generation models. Some of the minis were remote satellite terminals, three of them were destined to universities and the rest to specialized organizations, mainly places which were already using the Cuban minis CID 201A and B by the time they received the French minis.

On April 1969, the UH created the Digital Research Center (CID) with the mission of developing the Cuban computer. On 18 April of 1970, the first Cuban minicomputer, the CID 201, became operational. It was a third-generation model, with DTL integrated circuits, a 4K memory with 12-bit ferrite core words. Its architecture did not follow a full line of compatibility with the PDP 8. They connected a teletype to it with a paper tape I/O system at 33 bits/sec. In October of 1970, they also connected to it an audio compact cassette tape recorder at 300 bauds, ten times faster than the paper tape, and in addition, an auxiliary memory of up to 64K words by which its applicability, productivity, and reliability were largely increased [6]. This achievement occurred five years before reports of the first standards for similar use in microcomputers; it became an alternative for the expensive flexible disk units in use by then [7]. Eighteen units of the CID 201 were manufactured and delivered between the end of 1970 and 1972, under the name of CID 201A, a technological variant whose only difference relied on the casing.

The fact that the original DEC systems were inaccessible for Cuba and the lack of compatibility between the CID 201A and PDP 8 programs became an extraordinary challenge for the CID researchers. By the end of 1970, the local software of the CID 201 included a basic binary and octal input/output system, an arithmetic package of integer and floating point, a LEAL 201 compiler (Algorithmic Language – auto code of indigenous design in polish and symbolic notation, which followed some of the concepts of Autocode Elliot 803B). By 1972, they added an interpretive compiler of a simplified FORTRAN and some other programs. Additionally in 1972, the experimental CID 202 minicomputer went into service with its research objectives. It had a 16K ferrite memory, with 16 bits words, an original architecture with the novelty of running two simultaneous programs with their respective operators.

By the end of 1972 Cuba saw the completion of the new family of minicomputer CID 201B; however, it was not wholly compatible with the PDP 8. It doubled the 201A speed, with a memory unit of up to 32K words through 4K modules, eight auto index registers by module over the first eight address of each zero page, interruption system for the I/O devices and a bus for direct access magnetic disks and tapes. The computer accommodated various software packages that included an original powerful FORTRAN IV compiler. In October 1972 witnessed the definition of the next generation of minicomputers, the CID 300; it would be wholly compatible in software with the PDP 11, following DEC's more innovative tendencies.

In 1973, CID started a minicomputers factory in the industrial area of the V. I. Lenín Vocational School based on its own indigenous expertise. Students had full participation in the production lines as part of their training and vocational development. This practice proved fruitful for them as they continued their higher education studies as engineers or mathematicians with early deep and advanced theoretical and practical knowledge on computing.

To summarize, therefore, in 1970 UH launched its Computing Commission with a national reach. It strengthened and diversified its research program, incorporating others for the industrial development as well as the applications and fast development of the teaching of the computing disciplines in the national education system. In 1970 the degree in the Computing Sciences began at the School of Mathematics of the UH and at the Central University; numerous faculties and colleges introduced the subjects of FORTRAN IV Programming and Analysis. By 1971 the UH introduced the area of computing engineering for students who had finished their fourth year in telecommunication or in automatic controls. The CID strengthened its academic department as it created courses on operation, programming, and applications, starting a master degree program in digital systems in 1971, in conjunction with Canadian universities.

3   Main Characteristics and Accomplishments of Cooperation Work in EC between the Institutions of the FSU and Cuba (1972-1990)

3.1   EC Tendencies at the Beginning of the Socialist Collaboration

In 1969 the CMEA adopted the agreement to create the socialist central computers (CC - mainframes), five years after the IBM 360 generation of computers appeared. The conclusion of agreements for minicomputers occurred in 1974, nine and four years after the launching of the PDP 8 and PDP 11 respectively, when the microprocessor and integrated memories market had just been born, with a fast incorporation to EC. Local tendencies and needs reinforced the popularization and massive use of computing because it provided greater power at lesser costs. It was already an extraordinary challenge to reach the world upper levels; to stay abreast and to keep a reasonable follow-up and to prevent the relative backwardness from increasing was a very difficult goal.

3.2 First Steps in the EC Collaboration between the FSU and Cuba

Cuba joined the CMEA three years after the agreement to develop the CC systems of the Unified Systems of Electronic Computer Machines. The Cubans already had some essential scientific and industrial knowledge in order to consider higher-level objectives in EC, independently of its economy, which did not yet surpass the characteristics of a poor developing country. That is the reason the world classified Cuba as one of the least developed countries of the signing members.

In 1974 when they reached agreement for the development of the System of Mini Electronic Computer Machines (SM MCE), Cuba, represented by CID as a national head organization, was one of its founding countries [8].   That is how the collaboration with the Soviet INEUM started, the leading multilateral organization of the project SM.

With the creation of the National Institute of Automated Systems and Computing Techniques in 1976, a member of the Council of Ministers, the Cuban-FSU collaboration in EC was prioritized at a high national level, thereby strengthening and intensifying the relations and compromise among institutions and peoples of both countries [9]. The FSU signed the Cuban specialization in central processors, video terminals, keyboards for video terminals and microcomputers, as well as software products for the SM systems.

3.3 Main Projects and Joint Outcomes

The Cuban-Soviet projects were part of the minicomputer generation of the SM second architecture, because it included among its guidelines the full compatibility of programs with the PDP 11 family of computers [10]. This would facilitate a more effective contribution to Cubans, due to their experience and knowledge of the DEC lines. This accomplishment occurred despite the fact that the economic blockade, to which they added the CoCoM regulations among others, impeded the access to original products and documentation [11].

The CID participated in the SM 3 and SM 4 development, including the joint startup of the first prototypes of the SM 3 in 1976, carried out by INEUM. The Cuban variant of the SM 3, the CID 300/10, reached a conclusion in 1976; its national tests began in 1977 and in the international essay, they codified it as SM 2301, which was in serial production between 1977 and 1988 with a total amount of 410 systems. In addition, they created a Cuban prototype of the SM 4 called the CID 300/20. They also developed the decimal arithmetic processor for CID 2201 (SM 50/02).

From 1974 to 1990, Cuba developed numerous alphanumeric and graphic video terminals, supplying more than 12,000 units to Soviet institutions in addition to satisfying the Cuban needs [12]. Among the main objectives stated was the gathering of experience with the raster–scan technique, incorporated to the CID 702 (SM 7203) in 1975-1976. They used advanced controllers through analogs from Intel 8275 and 8279. The functional analogy with the most advanced models of leading marks, with solutions and their own resources constituted key objectives. The CID 7205 (SM 7213) emulated the DEC VT 52 and they incorporated a graphic regimen to this model. Other terminals emulated were the VT 100 and the VT 240 [13].

Between 1984 and 1990, various models of keyboards were developed and manufactured, fulfilling the SM standards and guaranteeing the functional analogy with those of IBM PC, VT 100, VT 240, and others. They used in abundance the contact key by electrically conductive rubber; Soviet institutions received dozens of thousands of keyboards from Cuba. The Cuban and Soviet specialists, who jointly developed these projects, reached a favored place among the keyboard and terminal video designers for SM.

With the software for SM, they obtained important results, among them the COBOL CID 300 compiler, tested for the SM in 1979, and widely used with Soviet minicomputers. Some other software products were the management system GES300, the multi terminal SMT300, the dBASE300, the UNIX tools and the COMUNIX CID300–IBM.

4 Conclusion

Nowadays, both the organizations and the Cuban specialists that worked closely together for more than eighteen years with their Soviet siblings are devoted to other development and application of modern informatics, where by some ways many of the results of that long and fruitful collaboration remain. In the 1980s, the ICID began a gradual redirection towards the development and production of high-level medical equipment and applications for diagnosis, intensive care, rehabilitation, and reaching a very high prestige and quality in that specialty. These equipment and applications are used profusely in the national health system and are exported to several countries. A similar process happened with complex automation, now specializing in applications for the sugar industry, the medical and the pharmaceutical industries, and the tourist facilities as well.

Certain factors, beyond the objectives of the current work, caused the advance and the level of the results of the socialist CE to lag behind in respect to the advance of leader countries at world levels. Nevertheless, in that period the collaborative work made unquestionable and everlasting contributions to human development. Among them are the contributions to scientific knowledge and the graduation of a great mass of highly qualified specialists, that were in those moments placed at the disposal of countless and diverse institutions. It is an irrefutable fact that as a result of this bilateral collaboration, at least modest contributions were added to those objectives, along with the friendship and mutual knowledge achieved among Cubans and formerly Soviet citizens, forever Russian citizens. The history stated here and much more are part of the history of computing of these countries.

References

1. See IEEE Santa Clara Valley Section, Dedication (May 26, 2005),
   http://www.ieee.org/organizations/history_center, reports that IBM
   305 RAMAC was commercial from (September 4, 1956)
2. More FAQ about IBM,
   http://www-03.ibm.com/ibm/history/documents/pdf/faq.pdf
3. Barquin, R.C.: The State of Computation in Cuba – DATAMATION, pp. 69–72
   (December 1973)
4. Sáenz, T., Capote, E.: Ciencia y tecnología en Cuba. La Habana, Editorial de Ciencias
   Sociales (1989)
5. López, J., Tomás: Cubans have got a special gift for mastering the computing – Juventud
   Rebelde Journal (March 23, 2006)
6. Ball-llovera, D., Antonio – CID report – 1970 and later documents (1970)
7. KCS (Kansas City Standard) o BYTE standard, BYTE February, 1976. A little later the
   Processor Technology Corporation published the popular CUTS – Computer Users' Tape
   Standard – which work at 300 or 1200 bauds,
   http://en.wikipedia.org/wiki/Kansas_City_standard
8. Egorov, G.A.: SM MCE. Detailed Scheme (July 2, 2002), http://www.computer-
   museum.ru/histussr/sm_evm2.php
9. Nitusov, A.: Computing Technique of the CMEA countries, Is a very interesting paper, but
   information about Cuba is not enough accuracy (November 2005),
   http://www.computer-museum.ru/histussr/sev_it.htm
10. Filinov, E.N.: System mini ECM (SM MCE), http://www.computer-
    museum.ru/histussr/sm_evm.htm
11. CoCom: Coordinating Committee for Multilateral Export Controls. CoCom was
    established during the Cold War to put an embargo on Western exports to East Bloc
    countries, http://en.wikipedia.org/wiki/CoCom
12. The XXX Anniversary of the creation of the first Cuban computer, the CID 201,
    publicized in CD by ICID (April 2002)
13. ICID – Instituto Central de Investigaciones Digitales, the new name of CID, modifying its
    category in the decade of the 80s (1969)

Teaching Computer Science in Moscow Universities: Evolution for Forty Years

Olga Parakhina[1] and Yuri Polak[2]

[1] Stankin Technical University, Moscow, Russia
parakhina@yandex.ru
[2] Moscow State University, Moscow, Russia
yuripolak@yahoo.com

Abstract. The authors share their experiences of teaching computer science at Moscow universities since early 1970s. During this period the technical basis, as well as content and methods of teaching have dramatically changed. The authors state that studying of modern information technologies promotes the creative development of students; and the best way to prepare students for further work is involving them to real research. Some examples of such collaboration of students with academic institutions and high-tech companies are given in the article.

Keywords: Computer education, history, hardware, information search.

1   Early Period of Computer Education

Education is critical to each citizen's ability to thrive in the knowledge economy. Today's students must develop key 21st century skills such as familiarity with problem solving, critical thinking and collaboration, and especially information and communication technology. But still half a century ago there were no preconditions for mass training to these disciplines. The quantity of computers was insignificant, they were bulky and expensive, and only skilled experts could communicate with them, using language of machine codes.

Authors (at least one of them) were taught programming since early 60s. Computer training of schoolchildren was spent then only in few special schools and special workshops in some colleges. As early as in 1960 the first graduation of schoolboys with qualification 'programmer' in the Moscow school #444 took place. First generation computers on electronic lamps ('Ural-1', M-20, 'Elliott') formed a technical base of training. The general-purpose electronic digital computer 'Ural-1' was manufactured in Penza in 1955-1961. It was designed for solving factory planning; accounting, statistical and other problems. Some of its characteristics appear in Table 1.

During training, schoolchildren carried out various educational tasks, e.g. solving systems of linear equations, matrix operations, functions tabulation etc. Later, during education in the Moscow State University (late 1960th-1970th) authors were engaged in developing algorithms and programming on Algol language. After graduation from the University authors have been conducting pedagogical work in several Moscow

J. Impagliazzo and E. Proydakov (Eds.): SoRuCom 2006, IFIP AICT 357, pp. 258–265, 2011.

colleges and universities (MGAPI/VZMI, MGIU/VTUZ-ZIL, MIREA, MESI, MATI etc) using various Soviet-made computers ('Dnepr', 'Minsk', 'Nairi', ES-1020 etc) and different programming languages (Algol, Cobol, Fortran, PL/1, YaAP). That time computer science education everywhere occurred without direct contact between trainees and computers as training system based on batch mode principles. Students gave blanks with texts of programs to their teacher and (after card punching) received results of its performance in a few days in the form of paper listings. Debugging procedure required repeating these operations again and again.

Table 1. Ural-1 Characteristics

computation speed	100 operations/sec
number base	binary
instruction type	single-address
number of instructions	29
storage capacity, 36-digit codes:	
magnetic drum	1024
punched tape	10000
magnetic tape	40000
speed of information output:	
to printing devices	100 lines/min
onto punched tape	150 codes/min
power consumption	10 kVA
floor space	75 sq.m

Fig. 1. 'Ural-1' in Moscow school #444; 1965[1]

Prior to the beginning of 1980th training to computer science was conducted only in natural-science, technical and economic colleges and universities. The basic disciplines were programming and algorithmic languages (usually Algol, Fortran or PL/1). That time computer literacy was understood exclusively as skill to

---

[1] Source: http://schools.keldysh.ru/sch444/01-04.htm

programming. Educational process had met a number of difficulties because of some problems (and not all of them are solved till this time). First one was the poor technical equipment. In the majority of educational institutions teaching of computer science was carried out then without any computers (there was a popular joke: teacher distributed clay to schoolboys and asked them to mould a computer). Secondly, that is not less important, lack of the qualified teachers, especially in secondary schools. At best, teachers of physics or mathematics taught computer science at schools. Therefore the learning efficiency was enough low.

2   Back to Dialogue

Computers have been separated from trainees for years. This situation was changed radically after arising of terminal classes for mainframes and especially desktop microcomputers. The opportunity of dialogue between student and machine has introduced absolutely new qualities into educational process. Acceleration of program debugging and controlling any stages of computational process became possible. Students' understanding of computer functioning had improved, and their interest to lessons had grown [1].

However, the technical equipment for these lessons remained outdated and inefficient. Soviet Union has catastrophically lagged behind the West in computer science and high technology, especially in telecommunication and microprocessor equipment. Some reasons promoted it, but the main were ideological and economic reasons. For many years in the USSR, there was a strict control over information dissemination and any printing or copying equipment. Authorities felt the latent threat in the idea of a personal computer, which assumed information freedom and human development. From the other hand, Soviet economy was extremely nonflexible because of state regulation, and USSR could not quickly organize manufacturing of personal computers. During 'cold war' import of many necessary components was impossible because of COCOM restrictions, so 'Motorola 68000' processor was under strict embargo.

That's why teachers of CS were compelled to use unproductive and inconvenient domestic computers. But even their amount was too far from real needs. For instance,

Table 2. D3-28 Characteristics

element base	K155, K565 microchips
computation speed	300 operations/sec
notation	binary, decimal, hexadecimal
RAM	16 Kb
ROM	4 K words
	built-in cassette tape recorder
power consumption	150 VA
optional periphery	typewriter, photo input reader, graph plotter, puncher etc
weight	24 kg

'Electronics D3-28' which was the main training means in MIPT, MADI, VZMI and other institutes in the beginning of 1980th, was made at three plants in total not more than 5000 / year. We remember our expectations within many months in queues of potential users directly at plants, so it was difficult for us to equip classes; it took 5 years to buy 60 computers (and monitors for them from other plant). But even such imperfect equipment has allowed to introduce dialogue methods and BASIC language into educational process. The 'D3-28' characteristics appear in Table 2.

Fig. 2. The class of 'D3-28' computers; 1980[2]

3 Post-'Perestroika' Development

In 1985, Mr. Gorbachov came to power, and democratization of Russia has begun. The same year obligatory studying of computer science at secondary schools was started. "The basics of informatics and computer science" course for 9-10-graders and teacher's manuals were published under academician A.Ershov's supervision; a methodical magazine 'Computer science and education' had appeared. Domestic educational computers (Agate, BK-0010, Corvette, DVK) formed a hardware basis of CS education. Private initiatives (cooperative societies, joint ventures) had provided import of the personal computer and development software for them (Cyrillic fonts, educational programs etc). The main goal was training pupils to programming and development of algorithmic thinking.

In higher education that period teaching of computer science had extended to humanitarian disciplines, but programming still remained as a main content of education. In 1990th intensive distribution IBM PC in Russia had attracted noticeable changing the content of CS courses at all steps of education. As mass preparation of new equipment users was required, many colleges have made computer science an obligatory subject at all faculties. They finished to teach programming, having passed to IT-training. The motto 'programming is the second literacy' was replaced with the new purpose, 'mass computer literacy'.

Besides hardware, huge changes in the agenda of training courses, in principles of laboratory practical work organization took place. Chalk and cleaning cloth were

---

[2] Source: http://asoiu.istu.ru/1_3.htm

replaced by felt-tip pens and markers, then notebooks and projectors. Now lectures are accompanied by multimedia demonstrations, and 1st rate students have skills of professional information search and use specialized software packages.

The higher education information infrastructure was improved as well. Federal university network RUNNet was created in 1994-96; followed by infrastructure for basic network for a science and education RBNet (1996-99). The same time Soros Foundation opened Internet-centers in 33 leading universities. Since 2000 the Federation of Internet Education created about 50 centers for training teachers to possess Internet-technologies. One more private initiative, Intel Teach to the Future Program, is a worldwide professional development effort to help teachers. Launched in 2000, the program has reached more than 3 million teachers in 30 countries (300 thousand in Russia). Through this free program, teachers learn from other teachers how, when and where to incorporate technology tools and resources into their lesson plans. Participants create assessment tools and align lessons with educational learning goals and standards.

4 Contemporary Problems of IT-Education

One of the main problems in IT-specialists training is a gap between plans of academic education and requirements to graduates, employed in companies. Colleges and universities curricula don't follow modern tendencies of branch, so companies should retrain their staff. It requires not less than half a year, or even 2-3 years for highly skilled employees' formation. In a number of universities there is nobody to teach the specialists of modern level, since many leading teachers and professors have retired, or left abroad, or work in other spheres, not according to their qualification. The average age of the faculty is close to 60 years, and their salary is less than underground cleaners' one.

The quantity of students becomes less as well. Alan Martinson has mentioned demography problems in his presentation during the Russian Outsourcing and Software Summit (ROSS-2006; http://www.soft-outsourcing.com). Here are some figures from his presentation. There are nearby 1.3 million people in the age of 17 in Russia now, but in ten years this quantity is to be decreased approximately up to 650 thousand person, and only then the curve should go upwards. According to Martinson, now Russian colleges graduate 40-50 thousand 'pure' IT-specialists annually, plus 200-250 thousand graduates have base knowledge in IT-technologies. But there are a number of new and actively demanded specializations where education in universities is not conducted at all, or the quantity of graduated experts is not enough. Among such specializations of scarce professionals are project managers; experts on testing; technical writers; manual developers; experts on interface ergonomics; analysts; experts on consulting and project implementation; and lawyers with IT-specialization.

One more essential feature of IT is the high speed of changes of applied technologies. The specialists should update their knowledge essentially at least every 5-6 years. Therefore it is necessary for Russian IT-education to establish close communication to labour market and to provide flexible reaction of education system to its inquiries, and also to develop system of lifelong education and personnel retraining. Russian IT-companies should play an important role in this process. They can help in definition of requirements to graduates knowledge and curricula; in launching the centers for additional and certificated training etc.

Many IT-firms actively cooperate with educational institutions, introducing joint programs of training the students of senior rates, so they are prepared for work in IT-industry on a student's bench. Computer business helps to introduce both the newest tendencies in the field of technologies and needs of branch into conservative curricula of higher education system. Here are some examples of such cooperation. The LANIT Company conducts joint activity with the Higher School of Economy, giving an opportunity to students to pass training, and teachers can raise their qualification on company's courses. One more project of LANIT is the computer school 'Expert' organized together with Computational Mathematics Faculty of Moscow State University. Undergraduates and post-graduate students of Moscow Institute for Physics and Technology can maintain their theses for master's degrees and Ph.D using their labour experience in the base company 'Phisicon'. In 2003 in the Nizhniy Novgorod University (under support of Intel Corporation) the Laboratory for Information technologies was created. Its mission is to form a stable system of education of highly skilled experts in modern computer technologies in view of IEEE-CS and ACM Computing Curricula recommendations for the enterprises of the information industry (naturally, for Intel as well).

In the St.-Petersburg State University chair of system programming is leaded by Prof. Andrew Terekhov who is a head of computer company 'Terkom' simultaneously. He considers that the best way to receive well-educated experts for the company is to train them necessary skills by himself. The teacher should be active researcher; he must tell students about his own research, instead of retelling another achievements. As Terekhov states, the cooperation between university and profile companies interested in carrying out of researches and students training is necessary. Such overlapping of practice and theory is successfully realized in St.-Petersburg State University. Education plans should correspond to international standards. Now only few Russian universities carry out requirements ACM/IEEE Computing Curricula (he estimates a level of his own university's conformity as 80 %).

The authors have similar experience as well. 10 years ago we have started to use students as researchers in real projects of Russian Academy of sciences. Simultaneously the CS curriculum in the Moscow State University has been modified in view of these practical problems.

5   On Teaching Computer Science at Moscow State University

As known, Internet has become not only communicative environment for scientific dialogue but also an invaluable source of information. Network technologies make researchers able to participate in international information exchange, while lack of Internet access rejects them to periphery of science. But sometimes unlucky searchers qualify Internet as 'information dump' (usually they ignore special tools and methods of effective search; don't use inquiries languages and search manuals; don't better their requests with additional keywords and so on). In Moscow State University we have developed a curriculum of computer science for sociologists with special attention to problems of network technologies and information search. Authors' course was changed dramatically four times for last 10 years. Each new version tries to meet the needs of the future research work and to approach the students to a level of competent users of the most widespread software. Now our course contains such sections as HTML documents, web design, etiquette of e-mail and – last but not least – methods of information search in Internet.

Studying of modern information technologies promotes the creative development of students. Some of them confidently work with networks, cooperating with the elder researches in search and analysis of information sources for humanitarian disciplines. The best students not only well use the existent search mechanisms, but also actively participate in development of new ones. Since 1996, tens of students of Moscow State University were working in the Central Economics & Mathematics Institute (CEMI) of the Russian Academy of Science as junior researchers. They participated in support of one of the best directories of Russian Internet resources which was introduced in 1995 in the Laboratory for Network information resources of CEMI (founded by Dr Yuri Polak). Its development required researching the mechanisms of search and making many original decisions; in particular, a special taxonomy (hierarchical set of rubrics) was created. Since 1996 our directory is available online; its origin was our database with more than 50,000 annotated records of information resources grouped on 300+ themes, which described both Russian sites and also relevant foreign materials [2]. Besides the standard search programs were used for the database updating. The students of MSU had brought a significant contribution to development of this unique catalogue of Runet (Russian Internet) by supporting a database and monitoring the information resources. They took active part in checking URLs, writing annotations and reviews. Many thematic reviews prepared by the students were published in well-known professional editions such as bimonthly magazine 'Russia's Information Resources', non-periodical 'Russian Internet Navigator' and PC Week (Russian Edition).

Our collective had won a prize in competition of young researchers of Academy of Science on fundamental and applied researches. Our students have also developed a number of well-designed web-sites with rich content, including unofficial pages of psychological department *flogiston.ru* and sociological department *www.nir.ru/socio/* of Moscow University. One more our project has been executed for the company Yandex, Russian portal #1. Its heart is a search engine with more than billion indexed documents of total volume about 30 TB. It has the largest database on Runet, vastly outstripping its nearest competitor. In early 2000s Yandex introduced a new advanced built-in catalogue. A group of MSU students took part in development of its taxonomies, database fulfilment and stuff training. The convergence of search engine and a catalogue allows overcoming their immanent defects: manually updated directories aren't enough full and actual; while the output of automatic search engine contains a lot of irrelevant links. This catalogue is structured specifically for the Russian market, showing web-sites based upon Yandex's 'Citation Index', a method of ranking web-resources by external references. This has proven very popular and the catalogue has quickly gained market share from the competition.

The amount and variety of Internet resources lead to complex taxonomies of universal Internet directories. As a possible solution to the problem, in Yandex catalogue a faceted classification was used. Facet analysis was developed as a tool for the organization of document collections in technical, scientific and social scientific fields, where it was highly effective in the storage and retrieval. Currently, facet analysis is used primarily to create classifications for the physical arrangement of documents (or document surrogates). In Yandex built-in directory such facets as 'genre', 'geographical region', 'source of information' and so on are used, so the search process become quicker and its results are more relevant.

The work in network environment expands mental outlook of the students, reduces time needed for fulfilment educational tasks, supplies them with unique information, improves their knowledge of foreign languages. Choosing the optimum decision of a search task, the students are being trained in network technologies and logical thought; develop their skills of information processing. These skills are necessary for their further research work and other activities [3-5].

We do not know how the world of information technologies will change in seven to ten years. However, we are sure that this progress will noticeably change our life and, as consequence, educational curricula. We will wait these new challenges with great interest and try to meet it adequately.

References

[1] Polak, Y.: Computerization of Education: Problems of Mass Learning. Problems of Economics 29(12) (1987)

[2] Polak, Y.: Internet in Russia, Russia in Internet. In: Proceedings of 22nd International Online Information Meeting 1998, London (1998)

[3] Polak, Y.: Information search in Russian Internet on the eve of 2000. In: 1st International Conference and Exhibition, Internet: Technologies and Services - IEEE (October 1999)

[4] Polak, Y.: Network technologies and creative development of students. In: International Conference TET 2001, Telecommunications for Education and Training, Praha (2001)

[5] Parakhina, O., Polak, Y.: Competitive and Game Components in Teaching Search Strategies. In: Proceedings of the 4th International Workshop on Computer Science and Information Technologies, CSIT 2002, Patras, Greece (2002)

Kronos: Processor Family for High-Level Languages

Dmitry N. Kuznetsov[1], Alexey E. Nedorya[2], Eugene V. Tarassov[3],
Vladimir E. Philippov[4], and Marina Ya Philippova[5]

[1] California, USA
leo.kuznetsov@gmail.com
[2] Veberi Ltd., Yaroslavl, Russia
a.nedoria@webaby.ru
[3] Intel (Wind River), Inc.
eugene@largest.net
[4] Ershov Institute of Information Systems, SB RAS, Novosibirsk, Russia
fil@iis.nsk.su
[5] xTech Ltd., Novosibirsk, Russia
flm@xtech.ru

Abstract. The article describes the history of the Kronos project that took place in the middle of the 1980s in Akademgorodok, Novosibirsk, and was devoted to creating the first Russian 32-bit processor, Kronos. The basic principles of high-level languages oriented architecture design are presented by the example of the Kronos processors family architecture. The article contains a short review of the original multiuser multitask operation system, Excelsior, that was designed and implemented within the project and used for the Kronos family processors.

Keywords: Kronos, Modula-2, 32-bit processors, high-level languages oriented architecture, OS Excelsior.

1 Background

Kronos is a generic name of a family of 32-bit processors for creating micro- and mini-computers. The Kronos processors architecture was designed to support high-level programming languages (C, Modula-2, Oberon, Pascal, Occam, etc.) that allowed implementing the newest ideas in software development and computer usage.

The Kronos family of processors was developed in the Novosibirsk Computing Center (SB AS USSR) within the framework of the MARS project ("Modulnye Asinkhronnye Razvivayemye Sistemy" – Modular Asynchronous Open Systems) by the Kronos research team of the START research and technology group (RTG) (1985-1988). The team leader was Dr. Vadim Kotov; the principal developers were Dmitry (Leo) Kuznetsov, Alexey Nedorya, Eugene Tarassov, and Vladimir Philippov. In 1990, START was transformed into the Institute of Informatics Systems named after A.P. Ershov SB RAS.[1]

---

[1] http://www.kronos.ru/

J. Impagliazzo and E. Proydakov (Eds.): SoRuCom 2006, IFIP AICT 357, pp. 266–272, 2011.

Kronos processors were manufactured as a pilot series mainly for building development computers designed for software development. The Kronos 2.6 processors were used for small-lot production of the Kronos-2.6WS Workstation. The first prototype of the workstation was demonstrated at the "Science-88" exhibition that was held in Moscow in 1988. Kronos-2.6WS workstations were used as development computers in a number of projects in the defense industry of the USSR. In particular, the Applied Mechanics Research used Kronos for developing satellite software and Production Association named after M.F. Reshetnyev (NPO PM, Krasnoyarsk-26).

One of the remaining Kronos-2.6WS workstations can be found in the Science Museum, London, UK, the Polytechnic Museum in Moscow, the Museum of Siberian Branch of the Russian Academy of Science (SB RAS, Novosibirsk), Novosibirsk State University (NGU), and IIS SB RAS in Novosibirsk.

2 The Kronos Project

The goal of the Kronos project was to create a multi-purpose 32-bit processor with hardware support of high-level languages. Kronos processors were used in open-architecture embedded microcomputers and single and multi processor workstations.

Modula-2, created by N. Wirth in 1975 based on Pascal (and later Oberon language), has been chosen as the main project language [1]. With the inheritance of the best qualities of Pascal, Modula-2 nevertheless had a number of distinguished features, making it most suitable for developing and implementing the software that fully met the requirements of its time. Such features included:

o Modularity (introduced the concept of module and possibility of its decomposition into definition and implementation parts);
o Well-developed data and control structures;
o Static type checking;
o Procedure types that allow dynamic parameterization of code by external actions;
o Low-level programming helper functions that made the control less rigid and enabled structural data memory mapping.

Architecture of Kronos is loosely based on the Lilith personal computer developed by N. Wirth at the ETH, Zurich. However, Kronos team made many independent decisions and moved to 32-bit architecture. For instance, while the sets of RAM value access commands, arithmetical and logical operations, and control structures remained unaltered, there were substantial changes in the architectures of process interaction, interrupt subsystem, addressing and working with peripheral devices. Many simplifications were achieved due to the 32-bit architecture.

The 32-bit word format allowed the use of the Kronos processors for solving computationally extensive problems. Implementation of some AI (Artificial Intelligence) prototype systems has been made possible due to huge 32-bit address space (up to 4 billion words). Hardware support of event interrupts, processes synchronization and inter-process communication as well as compact code made Kronos processors the preferred choice in some real-time automation systems.

268   D.N. Kuznetsov et al.

3 Kronos Processors Architecture

Kronos architecture was different from the traditional ones and had the following differentiating features:

1) Expression calculation took place on a fast hardware stack of small fixed depth, preserving stack content over function calls and switching process (threads);
2) The code and data of any processes were separated which resulted in ability to re-entry all programs and even their separate parts (modules);
3) Position independent code segment (PIC) with displacement table for starting a procedure addresses simplified the function call instructions and removed necessity for static linking;
4) Position independent data (PID) addressing scheme allowed a mapping of object-oriented concepts of modern programming languages. Independent base registers for local, global and external objects streamlined compilation process for high level languages;
5) Cross module base register addressing scheme enabled dynamic program loading – binding - execution;
6) Kronos architecture has been extended for operations with complex data structures.

To describe specific features of the Kronos processors architecture it is necessary to define some concepts. The program that is currently executed by a processor is a process. A process includes the following components:

a) Tables of currently downloaded code segments of separately compiled modules [DFT];
b) Global data field [G];
c) Code segment [F];
d) String constant fields [STRINGS];
e) Procedure stack (P-stack) [P];

In brackets [] there are base registers for the component address evaluation. Their meanings are as follows.

DFT (data frame table) – Supports dynamic binding of the separately compiled modules.
P – Stack has additional marking with the following indicators:
L – Base register that keeps the address of the beginning of the local data of the current procedure;
S – Base register that keeps address of the beginning of the free top part of the P-stack;
H – P-stack limiter (the limit of S extension).
P – Stack overlap (S >= H) is checked by hardware throwing interrupt exception.

The Kronos-architecture supports the following address schemes: local, global, external, intermediate and indirect. The values of all structural variables (arrays and records) were represented indirectly with reference to the beginning of the selected memory zone. Structural components of such values were located directly inside them. Thus, it was possible to limit the maximum size of a local data section of procedure to 256 words for fast addressing; that is, in terms of Modula-2 a procedure fast address first 256 local variables and takes more cpu cycles if the number of local variable is greater than 256.

Expressions were calculated on the register based fast arithmetic stack – the A-stack. The compiler statically traced the A-stack overlap and if necessary saved the content of A-stack to memory and restored it as needed.

M-code is a sequence of bytes without any bit tags. The first byte of the instruction determines how many (if any) bytes of immediate operand will follow. Most of the commands are in single-byte format. Extracting commands and operands from the sequence was carried out in parallel with incrementing PC (program counter) control register. M-code had four modifications of the length of an immediate operand: half-byte operand (when the last four command bits were interpreted as an operand), byte, 2-byte and 4-byte. Operands were represented by a hex, single byte, 2 bytes, or 4 bytes following the command code.

The command system was built in such a way that the most frequent operations had a shorter code, which increased the speed and reduced the code size. For instance, stack loading of the expression of the 0,1,...,15 numbers was executed by the LI0, LI1, ... , LI15, which were "load immediate" commands that together with the operand took one byte of the code. Similarly, the stack loading of the number 153 required the LIB, which was the "load immediate byte" command that together with the operand required two bytes of the code.

Procedures were represented with their own number from the 0...255 range. The code segment contained a "procedure displacements table" relative to procedures start. Therefore, the procedure call commands also had a length of one (for the procedures with 0...15 numbers) or two bytes. The mechanism of transferring parameters to the procedures varied. If a procedure had fewer than seven parameters, they were transferred via an arithmetic stack that reduced the costs of parameters processing in the call point. All arithmetic and logical operations worked on one or two upper stack elements and put the result on the top of the stack after popping up arguments.

The process descriptor at the beginning of the P-stack contains a snapshot of all registers at the moment of the last switching to/from a process. There is a process switching command that saves the current state of registers in the descriptor of a suspended process and restores the state of the registers of a resuming process, switching to it. This is similar to co-routine switching. It is very powerful way of hardware support for multi-tasking and synchronization. All interrupts were processed as process switching, indexed by interrupt number at the beginning of memory known as interrupt table.

4 Kronos 2.X Processor Family

The Kronos 2.X Processor Family included three models: 2.2, 2.5, and 2.6. All processors implemented the same system of commands. Processors differentiated by hardware implementation, cpu clock speed, and minor external devices variations.

The logic of functioning of all processor units was implemented in the micro-program control block. Two data buses connected arithmetic and logic unit, registers block, fast hardware seven words A-stack, command decoder, and communication buffer with the in-out bus. The two-bus internal processor structure enabled to execute binary operations on the stack (arithmetic, logical and other operations) in a single CPU cycle, which considerably increased processor efficiency. Micro-program control simplified processors and allowed implementation of complex commands. All processor hardware implementation is based on the Russian TTL (TTЛ) and TTLSh (TTЛШ) medium-level integrated components of widely adapted 155, 555, 531, 1802, 589, and 556 series.

The Kronos 2.2 was the first implementation of the above architecture concept. The processor was produced as a plug in board for "Elektronika-60" computer (similar to DEC PDP-11). The Kronos 2.2 was fully compatible with all devices that supported data communication protocol on Q-bus 22. The processor operated with 32-bit words. As the bus was 16-bits wide, the access to a word in memory required two Q-bus 22 cycles. The arithmetic and logic unit (ALU) for the Kronos 2.2 was 20-bit. Address arithmetic was processed in a single cycle, and data in two cycles. Such inner organization resulted in placing the whole processor on a single board using the element base of the time. The volume of directly addressed RAM of the processor was up to 4 MB; the clock rate was 4 MHz; the performance was 600 thousands operations per second on the stack arguments.

Unlike the 2.2 model, the Kronos 2.5 was fully 32-bits processor. It was produced in form of two boards. The interface with external devices was through the Multibus-1. Kronos 2.5 has up to 2 MB of local on-board memory. The CPU clock rate was 3 MHz; the performance was one million basic stack operations per second.

The Kronos 2.6 represented a further advancement of the 2.5 processor. It had an implementation and flexible reconfiguration designed to enable variety of usages. The processor could be embedded into a separate mini-computer as well as into a multi-processor complex.

The Kronos 2.6 was compatible with Euromekhanika bus and device specfication. The board size was 233.3 x 220 mm (E2). The basic set included the processing section board (ALU, stack, registers), a micro-program control board, a local memory board (min 512KB upto 2 MB), and the adapter board for the in/out bus. All devices were communicating via local synchronized 32-bit bus. The processor did not depend on a particular in/out bus. Compatibility with I/O bus was based on the specific set of external bus adaptors. Additional memory, inter-processor communication adaptor, local network controller, storage device on magnetic disks, code memory board (for separating code and data), bitmap-display, arithmetic calculator and other devices that expand processor capabilities has been implemented as a separate components on the local bus. The clock rate of the processor was 3 MHz; the performance was 1.5 million operations per second on stack operands.

5 Excelsior Operating System

Computers with the Kronos 2.2, 2.5, or 2.6 processors were controlled by the operating system Excelsior (OS Excelsior) that was designed as a general purpose operating system. In designing Excelsior OS, developers followed the basic principles of openness system interface, scalability, modularity, ease of integration, and user-friendly interface.

5.1 Open System Interface and Scalability

The OS design required to provide ways to accommodate wide variety of hardware configurations and allow the system to have only the components needed for specific usage scenario. Modular design of the OS and dynamic linking of all the components at run time accomplished it. The set of components presented at the particular installation of OS depended on the specific of the application.

5.2 Modularity

The qualities of Modula-2 programming language were the corner stone of the software modularity. In turn, software modularity created a basis for dynamic loading. Dynamic loading eliminated the stage static linking from the "edit - compile – launch" cycle and, therefore, considerably improved productivity of software engineers. Dynamic loading also meant there was no need to keep multiple copies of the same code – e.g. the frequently used modules, for instance, libraries. The absence of code redundancy, in its turn, saved external storage (in comparison with "linking" to the standard libraries in traditional systems).

5.3 Ease of Integration

Ease of integration was an attribute of all software tools of OS Excelsior that were built as a modular stratified hierarchical system. Practical application of Kronos computers as programmer's workstations in various projects showed that in most cases the developed software was not only independent software products but also sets of libraries that implemented major project abstractions. Those libraries were suitable for use with other software; they continually expanded the "construction kit" that Kronos programmers used for creating new systems.

5.4 User-Friendly Interface

Excelsior OS implemented user-friendly interface via well developed command line shell and binary file manipulation utilities. Kronos OS also included software engineer friendly screen text editor integrated with compiler and underlying shell execution environment.

6 Conclusion

Excelsior OS was based on Object Oriented ideas and abstract data types. Specific of devices were encapsulated in various implementations of driver modules that supported interface for all logical instances with a class of devices. Therefore, the system as well as application software communicated with devices only via well-specified logical interface.

Finally, it should be noted that all Excelsior OS software tools were oriented towards multi-process, multi-task, and multi-user applications. Due to its modular architecture, the Kronos processor addressed a wide range of applications, from embedded real-time systems to super-mini-computers. The most remarkable attributes of quality and programmability made Kronos indispensable in applications that required supporting continuous software development process.

7 Afterword

In September and October of 2005, Niklaus Wirth visited Russia. On September 21 at Moscow's Polytechnic Museum, Prof. Wirth has met Kronos research group (Vladimir and Marina Philippov, and Aleksey Nedorya). On behalf of the Institute of Informatics Systems SB RAN, V. Philippov presented to the Polytechnic Museum a functioning Kronos-2.6WS workstation. This workstation has been granted by Applied Mechanics Research and Production Association named after M.F. Reshetnyev (Krasnoyarsk-26 or, currently, Zheleznogorsk), where it was used in the 1990s as a space satellite software development workstation.

References

1. Kuznetsov, D., Nedorya, A., Osipov, A., Tarassov, E.: Processor Kronos in a multiprocessor system. Vychislitel'nye sistemy i programmnoe obespechenie, Comp. Center SB AS USSR, Novosibirsk, pp. 13–19 (1986) (In Russian)
2. Kuznetsov, D., Nedorya, A.: Design of symbol tables for languages with complicated visibility rules, Metody translyatsii i konstruirovaniya programm. Comp. Center SB AS USSR, Novosibirsk, pp. 153–158 (1984) (In Russian)
3. Kuznetsov, D., Nedorya, A.: Symbol files as the operating system interface. Informatika. Tekhnologicheskie aspekty. Comp. Center SB AS USSR, Novosibirsk, pp. 68–75 (1987) (In Russian)
4. Kuznetsov, D., Nedorya, A., Tarassov, E., Philippov, V.: Kronos, a workstation for a professional programmer. Avtomatizirovannoye rabochee mesto programmista. Novosibirsk (1988) (In Russian)
5. Kuznetsov, D., Nedorya, A., Tarassov, E., Philippov, V.: Kronos: high-level languages processor family. Mikroprocessornye sredstva i sistemy (1989) (In Russian)
6. Vasyokin, A., Kuznetsov, D., Nedorya, A., Tarassov, E.: Kronos: hard- and software support for high-level programming languages. In: Prelim. Proc. of All-Russian Methodological Conference on using Computers in Student Studies and Research, Novosibirsk, pp. 92–94 (1986) (In Russian)
7. Wirth, N.: Programming in Modula-2, 3rd corrected edn. Springer, Heidelberg (1985)
8. Wirth, N.: A Fast and Compact Compiler for Modula-2. ETH, Zurich (1985)

Author Index